The Gospel of Work and Money

POWER, POLITICS, AND THE WORLD

Series editors: Christopher R.W. Dietrich, Jennifer Mittelstadt, and Russell Rickford

Power, Politics, and the World showcases new stories in the fields of the history of U.S. foreign relations, international history, and transnational history. The series is motivated by a desire to pose innovative questions of power and hierarchy to the history of the United States and the world. Books published in the series examine a wide range of actors on local, national, and global scales, exploring how they imagined, enacted, or resisted political, cultural, social, economic, legal, and military authority.

A complete list of books in the series is available from the publisher.

THE GOSPEL OF WORK AND MONEY

Industrial Education and Its Global Legacies

Edited by Oli Charbonneau
and Karine V. Walther

PENN

UNIVERSITY OF PENNSYLVANIA PRESS

PHILADELPHIA

جامعة جورجتاون قطر
GEORGETOWN UNIVERSITY QATAR

Center *for* International *and* Regional Studies

Published with support from the Center for
International and Regional Studies (CIRS),
Georgetown University in Qatar.

Published by
University of Pennsylvania Press
Philadelphia, Pennsylvania 19104-4112
www.pennpress.org

Printed in the United States of America on acid-free paper
10 9 8 7 6 5 4 3 2 1

Cataloging-in-Publication record for this book
is available from the Library of Congress

Hardcover ISBN 978-1-5128-2893-1
eBook ISBN 978-1-5128-2894-8

CONTENTS

The Gospel of Work and Money

INTRODUCTION

Labor's Classroom

Oli Charbonneau and Karine V. Walther

In *The Souls of Black Folk* (1903), the pioneering African American sociologist W. E. B. Du Bois leveled a scathing critique against the educator and writer Booker T. Washington. Surveying the state of schooling for African Americans, Du Bois argued that Black leaders like Washington had prioritized manual instruction for people of color while denying them the broader liberal education accorded to white students in the United States. "This is an age of unusual economic development," Du Bois wrote, "and Mr. Washington's programme naturally takes an economic cast, becoming a *gospel of Work and Money* to such an extent as apparently almost completely to overshadow the higher aims of life."[1] In a time of intensified racial stratification, Du Bois feared an imbalance between a traditional liberal education and new forms of vocational training for nonwhites. What he elsewhere described as the "Tuskegee Machine" would enshrine "the alleged inferiority of the Negro races" through a permanent program of selective and limited skilling.[2] In Du Bois's eyes, the emancipatory rhetoric of Washington and his supporters masked a system that unwittingly aligned with the desires of white segregationists, maintaining forms of distinction and subordination that had emerged from racial slavery.

Born enslaved on a Virginia plantation in 1856, Washington bristled at this depiction of his work. He believed that Du Bois, who had been raised free in the North and schooled in elite institutions, did not understand the immediate needs of poor Southern Black people only a generation removed from enslavement.[3] During his youth, Washington was educated at the Hampton Institute, an industrial boarding school targeting African Americans and later educating Native American students. Although he never became a wage or agricultural laborer himself, Washington believed industrial schools could open a path for African Americans away from the worst degradations of the Southern white supremacist system.[4] "I believe most earnestly," Washington wrote in 1903, "that for years to come the education of the people of my race should be so directed that the greatest proportion of the mental strength of the masses will be brought to bear upon the every-day practical things of life."[5]

The debate over the "proper" form of education for African Americans rippled across Black intellectual discourse for decades. Marcus Garvey, founder of the United Negro Improvement Association, admired Washington's emphasis on bootstrapping economic productivity but broke with him on outcomes. "The entire Negro race must be emancipated from industrial bondage, peonage and serfdom," Garvey declared, viewing labor training as a means of asserting Black independence in a world increasingly ordered by capitalist production.[6]

Although remembered primarily through the prism of race relations in the United States, the Du Bois-Washington split channeled a globe-spanning set of nineteenth- and early twentieth-century debates about mass education and labor needs. These debates only grew in importance as the imperial nations that introduced labor education to their subordinated domestic populations began to export the practice to their overseas colonies. Du Bois's writings never explicitly tackled these exports, but he nonetheless recognized the links between the experiences of African Americans in the United States and the racial and economic justifications undergirding imperial expansion. In the same book where he leveled his critiques of Washington, Du Bois compared the economic benefits of both slavery and empire—and, by extension, the treatment of African Americans and colonial subjects overseas—and advanced the idea of what scholars would later call racial capitalism: "We have a right to inquire, as this enthusiasm for material advancement mounts to its height, if after all the industrial school is the final and sufficient answer in the training of the Negro race; and to ask gently, but in all sincerity, the ever-recurring query of the ages, Is not life more than meat, and the body more than raiment? And men ask this to-day all the more eagerly because of sinister signs in recent educational movements. The tendency is here, born of slavery and quickened to renewed life by the crazy imperialism of the day, to regard human beings as among the material resources of a land to be trained with an eye single to future dividends."[7]

Indeed, by the time Du Bois published his salvo in 1903, what had grown to be called "industrial education" had deep roots in numerous nations and had extended its reach throughout the colonized world. In many ways, Du Bois's critique of the "gospel of work and money" neatly encapsulates the nexus of education, capitalism, race, empire, and labor that defines our collection's central object of study: industrial education as a global practice with many afterlives.

Locating Industrial Education

The origins of modern industrial education programs arose from the European Enlightenment of the seventeenth and eighteenth centuries—initially in the writings of humanists seeking to build on the rationalist traditions of Gottfried Wilhelm Leibniz and René Descartes, and later in growing efforts to modify curricula and establish new schools that emphasized manual training. The emerging class stratifications of the period shaped these initiatives. While some advocates, like the philosopher John Locke, argued that practical skills improved the pedigree of young upper-class gentlemen, others focused their attention on the lower orders of society. An early eighteenth-century report from the Society of Sciences in Berlin argued that those "who are to take up a trade . . . [should receive] such training and instruction as is suited for their purposes and future condition in life."[8] From the outset, then, blueprints for what became industrial education often pivoted on the notion of improvement within the confines of social origin, something that would have powerful implications once these ideas spread. In British North America, colonial elites steeped in European literary culture drafted their own methods. President of Harvard College Leonard Hoar was an early adopter in the 1600s, advocating workspaces for students' "mechanic fancies."[9] In the eighteenth century, prominent figures like Benjamin Franklin, Benjamin Rush, and Thomas Jefferson built on Hoar's ideas, creating road maps for how general education might be synthesized with agricultural and industrial training.[10] While the term "industrial education" was not used, this early period enshrined connected ways of thinking about learning, labor, social and ethnic origins, and civilization that would underwrite the models explored in this volume.

By the latter half of the nineteenth century, most European and Euro-American educationalists agreed that modern (or modernizing) societies required pedagogies that emphasized "practical" instruction, even if a consensus on specifics remained elusive. This type of training was alternately described as "vocational" or "industrial."[11] "Vocational education" remains in use throughout the world today as a shorthand for directed forms of career training at secondary and college levels. Industrial education, however, has been remaindered—a historically obscure term that evokes, in a very general way, the mechanized sites of mass production that dominated global manufacturing by the early twentieth century.[12] Yet the term's roots are broader, deriving from the definition of industry itself—"the quality of regularly working hard"—rather than a specific connection to the factory worksite.[13] Industrial education grew from a capitalist moral order that valorized productivity and located output at the heart of individual and collective worth. It linked the functionality of the vocational—preparing students for a life of labor—with various forms of making: citizen-, nation-, empire-, and world-making. Industrial education was, in other words, a set of civilizational curricula designed to "modernize," "progress," and "develop." It spread globally through national, imperial, and corporate pathways and left deep imprints. Although initially advanced by the Western empires in service of an expanding capitalist world order, industrial education was malleable enough as a template to be adopted by state and nonstate actors with other ideological agendas. Its various adaptations would survive the decline and transformation of the imperial nations that birthed it.

The Gospel of Work and Money is a multidisciplinary exploration of industrial education and its global dimensions. Its chapters collectively argue that labor education—specifically forms promoted as "industrial education"—shaped nationally and imperially produced modernities, and provided templates for ostensibly post-imperial developmental ideologies. Our contemporary global condition, we argue, cannot be understood without a connective consideration of how these labor pedagogies and practices conditioned life on every inhabited continent. This volume represents the first significant attempt to bring new research on industrial education and its legacies from a variety of geographies and temporalities into shared discussion. In approaching this vast topic, we recognize that the industrial school and its curricula adopted a variety of guises—reformatories and prison ships, workshops and schoolhouses, mission stations and corporate plantations—and eventually shed the name "industrial" altogether. Advocates for industrial schooling were likewise varied: educational reformers, colonial officials, missionaries, anticolonial nationalists, development bureaucrats, and multinational corporations. Despite these qualifications, we can identify shared features across the following chapters. Industrial educators focused on manual labor while diminishing the role of abstract academic topics. They factored local environment and the broader economic needs of the polity into their approaches. They fixated on transformation through labor and its salutary disciplining effects. They coalesced around vulnerable populations—poor, racialized, criminalized, and movable—and promised stability or uplift. They deployed coercive tactics that reflected embedded power differentials around race, caste, class, and gender. They also fetishized visions of agricultural and industrial modernity and their "uplifting" potentialities. Industrial education practices undergirded a highly unequal world increasingly structured by capital and reliant on vast reservoirs of labor.

For much of the twentieth century, the scholarly literature on industrial education focused on how European and American educators developed manual training curricula within the context of state schooling for white children, writing about it as a prehistory of the increasingly standardized high school work shops and trade colleges of the twentieth century.[14] Absent in these early scholarly studies, however, was any sort of reckoning with industrial education's complex relationships with race, class, empire, incarceration, or the creation of exploitative labor regimes. By garbing the "manual arts" in the technocratic language of "scientific expertise" and "training," this literature obscured industrial education's key roles in disciplining labor in the metropole and producing workers for colonial regimes overseas.

In recent decades, successive waves of scholars studying labor and class histories, colonial empire, the history of education, and twentieth- and twenty-first-century development, among other topics, have challenged the depoliticization of industrial education, producing intimate accounts of how the practice embedded class in public school systems, fused with carceral reform, fueled the moral fantasies of empire, underwrote cultural genocide, tackled the "problem" of free Black labor, serviced the emergence of a global plantation belt, and provided rationales for postcolonial modernization initiatives. This new literature is often grounded in emergent theories of racial capitalism and empire as they relate to education, subjects that Du Bois himself tackled in his numerous writings.[15] Through internal colonialisms and overseas empire, racial capitalist modernity has deployed labor education as a means of enshrining inequalities in pursuit of hierarchized order and profit motives.[16] The global origins of this process are many and require deep historicization.

Befitting industrial education's scope, current literature on the subject crosses many disciplines and subfields within those disciplines. Despite this, scholars have not yet engaged in transdisciplinary and cross-historiographical collaborations that locate industrial education as a global phenomenon, instead working primarily within an array of national and imperial silos. While this type of framing is important, it can obscure productive comparative and connective dimensions of analysis, perpetuating the idea that industrial education has been a discrete—rather than a pervasive and organized—practice in the modern world. This framing elides the ways in which industrial education acted as a handmaiden to broad and frequently devastating structural and economic transformations that transcended borders. These included the global production of industrial and agricultural underclasses, the settler colonial domestication and monetization of territory, and the postcolonial integration of "developing" economies into the unequal vertices of the global marketplace.[17]

In pursuit of fuller understandings of industrial education and its legacies, the chapters in this volume explore multiple analytical seams. Chief among these is the history of childhood. The rise of mass societies and growth of colonial empires in the nineteenth century produced new ideas about childhood. These included debates over how to define it, chart its developmental phases, and improve it.[18] Industrial education arrived at a historical moment when child labor targeting marginalized populations—whether in metropole or colony—was prevalent. Rampant exploitation in the industrial slum, forms of bonded labor in the colonial hinterland, and the aftermaths of chattel slavery coalesced to inform new logics of reform connected to children's roles in society.[19] Crucially, these logics did not untether childhood from work, but rather reframed the connection. The worksite would no longer be the factory floor or the mine but rather the classroom itself, which would train children to be laboring adults, thus securing their value in increasingly capitalized societies. These gradated protections often did not materialize in practice. There were constant tensions, particularly in colonial settings, between capital's rapacious labor demands and the idealized (but in reality limited) safeguards afforded by industrial schooling.[20] Further, as several chapters in this collection demonstrate, practices of industrial education regularly overlapped with different forms of youth incarceration, in keeping with nineteenth- and early twentieth-century penological ideas about the connections between confinement and self-improvement.[21] The decline of the industrial education movement after the 1920s to 1930s saw the relocation of manual labor training for out-group populations to adult populations. Nevertheless, links to early education remained in the post-1945 period, notably in youth prisons and boarding schools.

Both intertwined and distinct, class and race also shaped practices and legacies of industrial education. The nineteenth-century acceleration of industrial schooling programs occurred in major part due to a perceived moral crisis generated by modern class systems. Political leaders and reformers in countries such as Germany, the United Kingdom, and the United States framed lower-class youths as parasitic and potentially destabilizing societal elements. Industrial education was a means of addressing this perceived social decay, resolving the threat of "idleness," and ostensibly rescuing poor children from exploitative labor conditions. The high-minded rhetoric of these programs arose from discourses about innate capacities and suitable life trajectories for the poor.[22] These ideas would be amplified by forms of race thinking, which developed and spread with the liberal empires of the latter nineteenth century. The relationship between race and industrial education was most evident in overseas colonies separated from the imperial metropole, but many national cores also deployed the approach to "manage" racialized populations.[23] While advocates of industrial education (especially missionary groups) positioned it as a free labor corrective to enslavement, in practice the industrial education programs of the late nineteenth and early twentieth centuries existed within the racial-hierarchical schemas of colonial societies, and thus cannot be disentangled from logics of exploitability targeting non-European populations.[24] As this volume shows, however, the relationship between race-based inequalities and industrial education was not always clear-cut. Racialized groups in both national and colonial settings used forms of manual instruction to stake legitimacy claims in a world structured by European-led modernities.

The classroom of industrial education was one demarcated by gender. Western-led global modernity brought with it ideas about what constituted male or female labor and, by extension, how these forms of labor should be taught. By the late eighteenth century, gender had become increasingly central in an emergent transnational "pedagogic discourse."[25] Growing from this, industrial education and allied educational forms sought to inculcate gender-"appropriate" monetizable skills from a young age. As several chapters illustrate, manual training for boys and young men often aligned with dominant models of masculinity, which emphasized physical labor in workshops, farms, and fields. Alternately, industrial education curricula assigned to girls frequently involved labor that could be performed in the home, whether one's own or that of an employer. In many nations and empires, monetizable skills like needlework, basket-weaving, cooking, cleaning, and childcare dovetailed with a desire to attune female labor to reproductive and domestic roles within the family. As Kirsten Kamphuis notes in her work on the Dutch East Indies, the use of "gender-specific education" within colonial settings was meant to "uplift" subject populations.[26] But this uplift was perennially grounded in gendered anxieties about the colonized body and its uses. Colonial industrial education was thus always in dialogue with broader concerns about native gender presentation, familial structure, and sexuality—sites of surveillance and control viewed by colonizers as foundational to the larger civilizing mission.[27]

Both secular and religious imperatives present in global modernity organized the racialized, classed, and gendered dimensions of industrial education programs and their successors. The conditions of possibility for industrial education arose from the fragmentation of "traditional" societies and their reorganization to meet the extractive needs of capitalist enterprise. Factory and plantation labor created the "problems" and "questions" industrial education sought to correct.[28] Advocates saw their resolution not in repudiating inequality-producing mass labor practices, but rather in reforming populations to interface with them more successfully. Industrial education, then, was both a response to and naturalization of the global labor reconfigurations occurring in the early national and late imperial periods.[29] Religious organizations were key to its initial spread, especially in colonies. This was particularly true of Protestant missionaries, who, according to Felicity Jensz, believed they were ideally suited "for providing indigenous peoples with Western, and implicitly religious, schooling" to shelter those same peoples against "the perceived negative effects and aspects of Western civilization."[30] If wage labor was inevitable, then learning to do it appropriately warded against extreme exploitation, modeled Christian modes of work and family life, and became a general means of advancement. This morality-wealth nexus blossomed under missionary supervision but became increasingly untethered from its Christian origins at the dawn of the 1900s, a century defined by the secular deification of modernization. As such, the worldview that Du Bois labeled the "gospel of work and money" was accessed and mobilized not only by those at the heights of capitalist, imperialist, and religious power, but also by marginalized historical actors, some of whom saw liberationist potential in building up labor capacity and directing it toward a variety of ideological ends, from radical anti-imperialism to nationalist traditionalism.

With these dimensions of industrial education in mind, the eleven chapters in this volume explore industrial education's varied manifestations and legacies in Asia, Africa, Latin America, Europe, the Middle East, and North America across the nineteenth, twentieth, and twenty-first centuries. Each case study is tied to a specific geography and emphasizes the transregional movement of people, things, and ideas that has produced—and continues to produce—forms of labor inequality. With this collection, we hope to broaden avenues of inquiry into the global practices of industrial education, moving away from curricular histories toward critical evaluations embedded in histories of race, class, gender, capitalism, and empire. This volume, then, is envisioned as a means of centering industrial education as an important historical phenomenon and encouraging new dialogues on its outgrowths and afterlives. The remainder of this introduction traces a global history of industrial education programs and considers how the chapters in *The Gospel of Work and Money* speak to one another and other geographies and temporalities of labor pedagogy.

Transatlantic Genesis

The massive social, economic, and political transformations Europe experienced between the early eighteenth and mid-nineteenth centuries produced the conditions necessary for industrial education to flourish. During this period, labor-based pedagogies transformed from experiments practiced in a limited number of German states to widely embraced approaches adopted by public and parochial educators in a growing number of countries.[31] The most influential figure in the drive for manual training was Johann Heinrich Pestalozzi, a Swiss educational reformer whose experiments in training poor youth on his own farm, Neuhof, in the 1770s helped shape some of the key tenets of industrial education. Pestalozzi's writings advocated experiential learning as a means of raising children up from poverty, making education economically self-supporting (often by selling handicrafts produced by students), and linking the traditional subjects of a liberal education to the labor of the workshop. His successors would translate these ideas into the industrial and agricultural schools of early nineteenth-century Europe.[32]

Adaptations of the Pestalozzian method catered to the needs of states and private enterprise alike, particularly in regions that had been reshaped by twinned industrial and urban expansions. Older apprenticeship systems, reliant on individual supervision and artisanal production, no longer seemed fit for purpose as populations swelled around factories, mines, and ports that required high volumes of pliant wage labor. As customary forms of vocational distinction were transformed into modern class systems, templates for mass manual education found more traction, first among reformers concerned with the poverty of Europe's unequal cityscapes and later within national governments seeking to standardize vocational training, create stable labor pools for employers, and ameliorate social strife created by the explosive growth of urban immiseration.[33] The state methods developed to address these issues in the mid-to-late nineteenth century differed by nation in important ways, with the market-driven British model standing in marked contrast to the centrally designed German one.[34] But they also shared features.

One of these features was a focus on what contemporaries considered "dead-end" children from the rural and industrial underclasses. In Great Britain, for instance, this elite preoccupation—held by moral reformers, politicians, educators, and industrialists alike—built on the long history of the Poor Laws, which, beginning in the late sixteenth century, attempted to alleviate poverty by confining or otherwise compelling impoverished Britons to work. Drawing from these laws, the seventeenth-century Houses of Correction designed apprenticeships for the "poorest children" that were frequently "disguised form[s] of servitude."[35] The issue of vagrant youth only grew during the First Industrial Revolution (1760s to 1840s), leading to the creation of schools promoting industrial and religious education in the late eighteenth century, which would later be followed by the "ragged school" movement in the 1840s designed to educate and "civilize" underclass children.[36] Ostensibly built to mitigate poverty and teach self-sufficiency, industrial schools also functioned as alternatives to the workhouse or jail cell—the beginnings of the durable links between labor education and confinement explored throughout this collection. As Christine Whyte details in her chapter on the imperial and carceral dimensions of industrial schooling in Scotland, a wave of acts in the 1850s and 1860s in the United Kingdom created new provisions for educating and containing poor youth in government-"certified" schools, which at their height numbered nearly three hundred. Reformer and governmental support for the certified school indicated a growing consensus in middle- and upper-class Britain that industrial education could prevent "vulnerable" children from transforming into parasitic criminals.[37] Foreshadowing the rhetoric later applied to colonial subjects abroad, the British Christian reformer Lord Ashley, Earl of Shaftesbury, described the potential dangers of the laboring classes: "thousands may be in our highways and hedges, in our streets and alleys, in our courts and lanes, who are living in a state of practical heathenism—a heathenism as complete as if they were found in California or Timbuctoo."[38]

In line with these concerns, a range of industrial education programs appeared in Europe and the Americas. In Western and Central Europe, their aims reflected the social challenges of industrialization and catered to the requirements of capital. Food scarcity and population growth in eighteenth-century Hamburg, for example, created "a new awareness of the need to reform the traditional system of poor relief" and increased investment into *Armenschulwesen* (schools for the poor) that prioritized the teaching of employable skills.[39] Swedish governmental support for vocational schools saw as many as ten thousand students enrolled in them by the end of the nineteenth century.[40] In Southern and Eastern Europe and in the postcolonial Latin American states, manual training began later in the century and had an aspirational character grounded in the desire to transform agricultural economies into industrial ones. Indeed, as Helge Wendt illustrates in this volume, countries like Chile and Mexico tapped transnational educational expertise to develop their industrial schools, with European and European-trained teachers translating textbooks and shaping curricula. In Eastern Europe, the Romanian legislature passed a vocational education law in 1899 that stipulated the establishment of schools to bolster state-run workshops producing tobacco, weapons and gunpowder, military uniforms, leather, and matches. They also placed special emphasis on orphaned or abandoned girls, who were thought to "pose a social danger," and trained them as servants, nannies, nurses, and teachers.[41]

The influence of Pestalozzi and his successors reached American shores in the first decades of the nineteenth century. By the 1820s, industrial schools had opened in Maine, Upstate New York, and Indiana. Among the most influential were those founded by William Maclure. A partner of the Welsh utopian philanthropist Robert Owen, Maclure established the Orphans' Manual Training School in 1827 and the School of Industry the following year, both located in New Harmony, Indiana. The goals of both schools were to "[fit] the pupils for useful occupations" and minimize the cost of their education "through the products of their labor."[42] Forms of manual training seeped into training at theological seminaries and literary institutions, with a particular emphasis on assisting the "'worthy poor' who wanted an education." In the 1830s, these could be found throughout the Northeastern states. Female seminaries included "manual labor as a means to inculcate industriousness and to help defray the expenses of the school."[43]

By the mid-nineteenth century, calls for systemized manual instruction in schools were common, although it was not until the 1870s that the practicable means to accomplish this emerged in the United States. Curiously, the broader push for manual education was also driven by enthusiasm among American educators for the imperial Russian system of manual training demonstrated at the 1876 Centennial Exposition in Philadelphia. But it also built on a century of evolving ideas about labor and education in industrializing societies on both sides of the Atlantic, and often occurred in parallel with the push for manual education among colonized subjects.[44] The desire among educators to see these systems grow coalesced with the dramatic economic and demographic growth of the United States between the 1860s and 1910s, which created many of the same driving factors (urbanization and large laboring underclasses) found in Northern Europe during the previous century.[45] With expansions of state-run education in the latter half of the nineteenth century, forms of manual training were poised to find wider implementation—a pattern that was occurring across the Global North.

Empire's Laborers

Colonial variants of industrial instruction co-evolved with the national models described above, albeit more haltingly and primarily through missionary initiatives rather than public or experimental schools. The liberal empires of the 1800s—built on a mutually dependent triad of resource extraction, human exploitation, and globalized commerce—increasingly trafficked in ideas of cultural reinvention along European lines as the century progressed. This was in large part the by-product of transatlantic Christian reformers, who pushed narratives about the responsibility of colonial authorities to introduce Western customs to their wards, but also served practical state goals of social compliance and labor creation. Educational programs featured in every empire of the era, although they differed in scope, spread, and uptake. French and Portuguese colonials in Africa and Southeast Asia engaged in projects of absorption, using education as a means of creating assimilated-yet-subordinate subjects, while education in the Dutch East Indies was systemically underfunded due to fears of Islamic infiltration and struggled to gain a foothold.[46] Government-backed British mission schools in West and Southern Africa packaged religious conversion and vocational training together as a means of spiritually and materially transforming native societies.[47] Being a "serious" empire meant introducing cultural and educational policies that serviced both the pretensions of the civilizing mission and the economic needs of the colony.

Race underpinned the construction of colonial industrial education, infusing the metropolitan binary of "deserving" and "undeserving" poor with more rigid (but often contradictory) ideas about ethnocultural difference and civilizational benchmarks. By training colonized children, reformers contended, missionary and state schools fulfilled a moral purpose, ensuring that native societies became capable of progress and escaped extinction. The industrial school served multiple ends for colonizers: as a vehicle for moral uplift and religious conversion; as a means of validating and perpetuating empire; and as a mechanism for individual and collective economic improvement.[48] Once established, training for youth varied based on the extractive potential of the environment, the aims and directives of individual missionary organizations and colonial governments, the perceived level of preexisting "progress" within a colony, and the degree to which incentive or threat elicited local participation in education.

Nineteenth-century missionary organizations were frequently the "initial and exclusive agents of education for indigenous and non-European peoples" in colonized and semi-colonized territories.[49] As the chapters by Bronwen Everill and Arun Kumar demonstrate in this volume, this was especially true in the British Empire, where religious societies dabbled in both metropolitan and colonial education. These policies, however, were also adopted by the French, Dutch, Portuguese, U.S., and Spanish in their own empires. Often critical of colonial authorities, missionary organizations maintained their own administrative structures, allowing them to work between multiple empires—thus ensuring their pedagogies did not remain sequestered.[50] The "major goal" of most institutions was the "creation of indigenous and local workers who were expected to contribute to the colonial apparatus."[51] While the French and Portuguese were more closely linked to assimilationist projects in their Asian and African empires, they too had missionary agents convinced that the "inculcation of working habits" deserved centering.[52] Even among Catholic missionaries, who tended to focus more on conversion and teacher training, a vocational emphasis seeped in by century's close. Plans for a boarding school on Java drafted by the Jesuit Frans van Lith in the 1890s included gendered labor training, something that was also present in schools run by the Franciscan sisters on the island.[53] Fraternal orders in the Spanish Philippines dabbled in rudimentary industrial and agricultural training from the seventeenth century onward, long anticipating its nineteenth-century implementation in colonial public schools.[54]

In the pre- and post-revolutionary United States, Protestant missionaries advocated boarding schools for Indigenous students that had large vocational components. Moor's Charity School in Lebanon, Connecticut, dedicated itself entirely to training Native missionaries and schoolteachers as a means of proving the viability of the Anglo-settler project to the British imperial metropole and extending Euro-American influence into the contested interior. Alongside language study and Bible lessons, the Delaware and Mohegan boys and girls learned how to mimic settler productivity through vocational and domestic labors.[55] The practice escalated after the American Revolution. By the 1820s, federal government allocations underwrote a variety of similar initiatives. Institutions like the Cornwall Academy and the Choctaw Academy offered a combination of language, religious, and vocational instruction.[56]

These schools anticipated the systematic use of industrial education as a means of solving the "Indian Problem" in the post–Civil War era. In 1873, under President Ulysses S. Grant's "peace policy," the government's approach to Native Americans, including their schooling, was largely turned over to Protestant missionary organizations, in what scholar Francis Prucha has argued "might just as properly have been labeled the religious policy."[57] The Secretary of the Interior's report that year included a subsection entitled "Manual Labor Schools," which maintained that "upon no other subject or branch of the Indian service is there such entire agreement of opinion from all agents and persons, connected directly and indirectly with Indian civilization, as upon the necessity of labor schools for Indian children. It is manifest that barbarism can be cured only by education."[58] As Janne Lahti demonstrates in his chapter, between the 1870s and 1910s, the Carlisle Indian Industrial School joined scores of other state-subsidized boarding and day schools in placing manual training at the heart of its curriculum, using it as a tool to quicken cultural erasure and state consolidation in the contested spaces of the American West.[59]

American missionaries in Asia, Africa, and the South Pacific were in constant dialogue with their counterparts in North America on issues of education. The foundation of the American Board of Commissioners for Foreign Missions in 1810 expanded this missionary reach, initially following pathways created by British imperialism before branching into the United States' growing settler domains.[60] The Hilo Boarding School on the island of Hawai'i, founded in 1835 by Reverend David Lyman, blended religious instruction with vigorous "physical training" meant to inculcate the imagined values of Anglo-Protestant civilization (and offset the costs of operating the school). Samuel Chapman Armstrong, the architect of industrial education for African Americans in the decades following emancipation, was born in Hawai'i, where his missionary father began opening manual labor schools in the 1830s and later served as the minister of education for the islands. Following the Civil War, Armstrong secured a government grant that allowed him to open the Hampton Agricultural and Industrial School, which modeled itself on Hilo and other Hawai'ian schools committed to spreading the gospel of work.[61] Hampton would become influential as a template for the Native American industrial schools, and, after 1898, in the enlarged overseas empire of the United States. It was under Armstrong's tutelage at the school that Booker T. Washington developed his ideas about industrial schooling, eventually leading him to found the Tuskegee School, which received funding and support from white industrialists looking to capitalize on Black labor.[62]

Vocational training for racialized out-groups appeared in multiple settler societies, providing the veneer of liberal humanism and paternal oversight for colonizers eager to eliminate heterogeneous indigenous lifeways not anchored in capitalist accumulation. Writing on Japanese Hokkaido, Katsuya Hirano describes a logic recognizable in other settler geographies: "settler colonization would *both* destroy and rescue Ainu from the consequences of colonization's own destructive acts."[63] On Hokkaido, education for the Ainu—chiefly Japanese language learning and instruction on agricultural labor—represented "rescue" from the ecological and sociocultural impacts of Japanese migration.

In Australia and Canada, the rise of manual training programs occurred in tandem with swelling settler numbers and analogous Indigenous territorial decline, undermining the notion that training for farm or factory mitigated the spatial and cultural eliminationism inherent in settler permanency.[64] Rudimentary forms of industrial education appeared in Australia as early as the 1830s. In the following decades, settler advocates pitched it as a "humanitarian intervention" that would serve Indigenous peoples by aligning them with the ideological and material goals of the colonial state.[65] Following the 1879 Davin Report, an investigation into Indigenous education commissioned by the Prime Minister, John A. MacDonald, Canada's culturally genocidal residential school system moved increasingly toward forms of industrial education, with advocates likewise positioning it as a form of protection.[66]

Interpretations of industrial education varied throughout the settler and extractive colonies of Africa and Asia, as did the degree to which it featured in missionary and public schooling. In German Southwest Africa, for instance, settlers required pliant labor, and thus manual training in Indigenous schools was common. This was not the case in German East Africa, where the first governor of the colony, Julius Sodden, recognized the need for "literate local staff" and worried about relying on schools run by foreign missionaries.[67] Scholars have also questioned the degree to which missionaries, the main vehicles for industrial education, had "firm notions" of its substance. Most, according to Andrew Barnes, defined it through what it was not: the "humanistic education" they themselves had received.[68] Given these backgrounds, the degree to which classically educated teachers could impart technical knowledge was questionable, meaning skills-based training often remained at a basic level. Further, the biggest "success" of certain schools was in using child labor to underwrite their own operations, rather than training students in employable skills. Such was the case in British India, where commentators noted the "failure" of "industrial instruction" to instill self-sufficiency in orphans.[69] These inconsistencies in industrial education methods throughout the colonized world inadvertently reveal its omnipresence. By the early twentieth century, wherever there was a colony there was likely a set of ideas about how to educate the Natives to work.

The growth of transnational and transimperial professional networks played an important role in the spread of industrial education throughout colonized Africa, Asia, North America, and Oceania. Among seasoned missionary educators, conferencing provided a means of sharing pedagogical techniques, developing comparative frameworks, and building links with secular educators and government bodies. Conferences proved especially important in the Anglo-American world. Among other topics, the 1910 Edinburgh Missionary Conference focused on "the three principles of industrial teaching . . . being adaptive to local conditions, providing skills for individuals to become self-supporting, and instilling dignity into labour."[70] The yearly Lake Mohonk conferences in Upstate New York, held from the early 1880s until 1916, featured missionaries, colonial officials, and government administrators exchanging ideas about how to reform settler and overseas territories through labor.[71] These circulations of "expert" knowledge also occurred outside formalized settings. Colonies often inherited ideas about education through preexisting institutions or local pedagogical strategies. Such was the case in Southwest Africa, where German settlers tapped into "industrial skills" taught by British and Finnish missionaries to "educate" the Nama and Herero peoples to become good laborers.[72]

The appeal of industrial education meant that it was constantly refined and recirculated. The experimental farms and training programs of the American West crept into the work former U.S. Commissioner of Agriculture Horace Capron did with Japanese settler-farmers, while settler enthusiasm in Zimbabwe for "educational methods that had been developed for the freed slaves in America after the Civil War" increased missionary and government investment in industrial education.[73] The carceral-educational nexus also traveled beyond Europe and North America. As Sarah Steinbock-Pratt argues in her chapter, the links between educating Filipino children and reforming Filipino prisoners in the American colonial Philippines carried on practices that traced their origins to the "ragged" reformatory schools of eighteenth- and nineteenth-century Great Britain.

In the Greater Middle East, industrial education became a means of both subduing populations in the wake of Ottoman imperial collapse and asserting national sovereignties in a global system still dominated by the Western empires. As one scholar writing for the *American Political Science Review* noted in 1927 about Iraq: "Many observers think that attention should be given to agricultural and industrial training rather than to liberal education," noting "a fear that a plethora of educated Arabs with little opportunity to use their learning except in government will cause undesirable political agitation."[74]

Nationalist movements in colonial, postcolonial, and semicolonized societies tapped into labor education as a perceived means of diminishing both internal dissent and outside domination, amplifying their legitimacy internationally. Helge Wendt's chapter on Latin American nations, for instance, illustrates how these countries, free from the yoke of the Iberian empires yet conscious of their subordinate status in a system constructed by the North Atlantic powers, used industrial education to expand national "expectation horizons" and compete more vigorously in the global capitalist contest. In nineteenth-century Siam, King Chulalongkorn used education as a means of protecting his own country's independence "by showing a modern face to the Western empires," a practice that, by the 1910s, led to experiments with vocational training.[75] As Dolf-Alexander Neuhaus argues in his chapter, for Korean nationalists caught between the Japanese and European colonial empires, manual training in schools was a potential tool of liberation, creating a break from traditionalism and strengthening Korea's case for national independence. Here, industrial education was envisioned as a tool of nationalist armor rather than one of colonial control.

In the late nineteenth and early twentieth centuries, private corporations and philanthropic entities increasingly joined nations and empires in seeking to expand support for industrial education worldwide. Their participation illustrated the intersections between racialized ideas about labor potential and capital's voracious need for workers. After the American Civil War, industrial education for Southern Blacks increasingly drew funding from Northern industrial philanthropists who sought to profit from emancipation by producing new sources of compliant wage labor.[76] The simultaneous global expansion of colonial resource extraction and production saw companies and private philanthropic funds invest in their own forms of privatized industrial education abroad. The Phelps-Stokes family, who had previously supported the American Colonization Society's efforts to relocate freed African Americans to Liberia, began investing in industrial schooling across Africa while simultaneously supporting industrial schools for Native Americans and African Americans in the U.S. "homeland." In Liberia, as Bronwen Everill's and Hossein Ayazi's chapters demonstrate, industrial schools received the encouragement of the Firestone corporation, which viewed industrial schooling as an ideal way of obtaining their own future source of malleable labor. The influence of these efforts led to the term "Phelps-Stokism," used to describe the export of race-based education programs from the Global North to the Global South.[77]

Global Legacies

Anticolonial nationalisms and transnationalisms blossomed in interwar and post–World War II Asian and African states and among Indigenous populations in settler societies. These movements featured ideologically heterodox groups of political actors—from anarchists to anticommunist nationalists—who increasingly rejected the racial hierarchies of empire and offered liberationist counternarratives of life after imperialism. Anticolonial mobilizations produced anxious escalations of imperial surveillance and counter-dissent operations, but ultimately the politics of the colonized, the massive disruptions to international and interimperial order caused by World War II, and the rhetorical devaluation of traditional forms of colonial rule led to cascading decolonization in the 1950s and 60s.[78] The collapsing foundations of colonialism threatened a host of imperial projects, not least of which was education through labor.

The survival of what was no longer called "industrial education" came through the racial capitalism of an ostensibly postimperial era, where major Global North powers and power blocs—the United States, Great Britain, and the European Economic Community—increasingly sought to transmute colonial development into national development. In the case of the United States, this meant the continuation of an "anti-imperial" model based on the spread of capitalist modernity to "underdeveloped" former colonies. This model sought to empower counterrevolutionary native elites who were fundamentally receptive to both U.S. global supremacy and capitalism, and willing to open their nations' agricultural, mineralogical, and aqueous resources to exploitation by Western capital. Nested within the broad modernization-developmentalist initiatives of the postwar era were training regimes that advanced a combination of aid, education, and corporatism that had emerged in the first half of the twentieth century. The goal of conditioning future workers for industrial capitalism remained a central mission of development programs as governments, nongovernmental organizations, private companies, and Bretton Woods institutions such as the International Monetary Fund and the World Bank engaged in and promoted educational reform—shaping the local rhetoric of the schools themselves.[79]

In some cases, these "new" education programs barely effaced their roots in racialized modes of industrial education. As Hossein Ayazi demonstrates in his chapter on the Liberia-based Booker T. Washington Agricultural and Industrial Institute, the developmental approaches adopted by American aid programs relied heavily on the practice of Tuskegee-era missionary- and philanthropist-backed industrial education. Increasingly, they bequeathed its legacies to the corporate concerns that grew Liberia's plantation infrastructure and thus had vested interests in creating and controlling labor. Across the globe, development specialists located "underdeveloped" peoples—found most often in previously colonized spaces with "limited industrial and market infrastructures"—and relied heavily on training programs as "disciplining . . . procedures of capitalism" for them.[80] Galvanized by President Harry Truman's Point Four Program and its successors, government aid and philanthropic organizations, such as the Carnegie Corporation and the Rockefeller Foundation, understood their mission through a Cold War and, often, Christian lens, wherein forms of vocational training (now advanced beyond the limited purview of nineteenth-century industrial education) would help win the battle against global "atheistic communism." In newly independent Global South countries from Nigeria to India, local elites chafed at the prospect of continued Western hegemony but selectively embraced the notion, enshrined in industrial education since the early nineteenth century, that the dignity of the individual and the competitiveness of the nation rested in an ability to serve the labor demands of capital.[81]

The Angloworld settler schools where industrial education once flourished changed over time—but striking continuities remained. In the United States, Black demands for full political rights and equality, including in the realm of education, saw the gradual transformation of industrial schools into full-fledged universities. Yet the "industrial school" survived in the segregated American South as prisons for African American youth and adults alike. The most notorious of these institutions, the Alabama Industrial School for Negro Children, still functions as a juvenile detention site today.[82] Ongoing coercive labor systems in American prisons, often justified as reformative and educational, offer another testament to the ongoing legacy of industrial education and its disproportionate impact on African Americans and other minority populations in the United States.

Gains made by Indigenous activists during the 1930s—the years of the "Indian New Deal"—were contested by a postwar conservative turn on Native American education within Congress and the Bureau of Indian Affairs, which returned to an emphasis on off-reservation boarding schools and assimilation policies.[83] In Australia, "stolen generations" of Aboriginal and Torres Islander children faced removal and assimilation-through-education coordinated by the Australian government well into the 1970s, while in Canada, the last Indigenous residential school run by the Department of Indian Affairs did not close until 1997.[84] The deadly impact of these schools on Indigenous children due to malnutrition, disease, and physical and sexual abuse has only recently gained popular media and governmental attention. While calls by Indigenous communities for greater action and redress have resulted in official apologies by political leaders, alongside the formation of truth and reconciliation committees, such examples of moral reckoning remain rare.

Cold War socialist development models—whether emanating from the Soviet Union, allied Central and Eastern European states, or China—positioned themselves as egalitarian alternatives to the colonial legacies of the West, actively attacking American and Western Europe-led modernization as updated iterations of capitalist imperialism.[85] Yet labor pedagogy was inherent in the Soviet communist project, finding expression in the interwar years as Leninist and Stalinist authorities attempted to solve the "nationality question" in the Soviet Socialist Republic rimlands and evident in model workers' cities like Magnitogorsk in the Soviet Union and Nowa Huta in Poland. Teaching people to work toward a utopian socialist modernity was at the heart of these projects.[86] As Lukas Allemann argues in his chapter on education for Sámi children in the Kola peninsula, efforts by the capitalist powers to counter communism through aid for racialized and/or minoritized groups were mirrored by the Soviet Union in the 1960s. Soviet authorities sought to introduce labor education to minoritized populations to better integrate them into the workers' state. Allemann's chapter, alongside several others in this collection, blurs demarcations between "East" and "West" and "colonial" and "national," and illustrates how states across the world deployed labor pedagogies to advance their agendas.[87]

The naturalization of labor education as a central pillar of modernity made it a constant in efforts to incorporate the Global South into capitalist and communist worlds *and* to reconcile poor and racialized populations within Global North nations. In the aftermath of the Cold War, Western labor education became a normative feature of state-building, wrapped up in the broader depoliticization of liberal capitalist models—now standardized as the only viable templates for national progress. If the creation and movement of capital and labor were both inevitable and desirable, there was little need to interrogate historical origin stories.[88] The final chapters in this volume complicate that narrative by exploring labor education and migration in the Middle East and South Asia. As Danya Al-Saleh, Zahra Babar, and Mishal Khan demonstrate, states within the Gulf Cooperation Council have worked closely with sending countries in South and Southeast Asia to ensure vocational training before worker arrival. These policies echo earlier historical efforts to shape and discipline laborers, seeking their subordinate integration within specific nodes of global capitalism. The national and ethnic identities of these workers in many ways define the types of labor for which they are selected and trained, reminding us that connected legacies of globalized industrial education and racial capitalism are rarely far from view.

* * *

In 1846, the *Annual Report of the Poor Law Commissioners for England and Wales* described refinements to a system that was at once punitive and pedagogical: a sprawling matrix of workhouses and asylums for "vagrants" and "lunatic poor," aimed at converting industrialization's surplus populations into disposable free labor, confined prison labor, or settler workers in the overseas colonies. It reserved a section for Ireland, a colonized zone of exception considered racially distinct from Anglo-Saxon Britain.[89] Reports like this found voice for their concerns in what became known as industrial education, which would echo across the modern world. Its central preoccupations animated government ministers and colonial bureaucrats, schoolteachers and missionaries, corporate philanthropists and anticolonial nationalists. In time, as our volume argues, labor training established itself as a key component of civilization or development (depending on the era within which one wrote), and addressed concerns about national and imperial progress; integrated the labor demands of the modern state and private enterprise; revealed the intimate proximities of the workhouse, prison, and classroom; and then naturalized itself within templates for late-capitalist modernity.

The eleven chapters in *The Gospel of Work and Money* parse these histories of industrial education and explore their adaptations and legacies. Collectively, the contributors demonstrate how ideas about race, class, and gender intersected with capital, nation, and empire to shape labor education. In doing so, they position labor pedagogies as a constant feature of global modernity implicated in myriad processes of national and imperial formation. Their scope is wide: from prison ships off the coast of Scotland; to Philippine classrooms under U.S. colonialism; to the trade colleges of modernizing Latin American nations; to Japanese colonial training in Korea; to corporate plantations in West Africa; to Soviet residential schools; to labor recruitment centers in contemporary South Asia. Unsurprisingly, a remarkable degree of variability exists within these case studies. As the chapters show, industrial education and its successor forms often had intentions and outcomes specific to the national and imperial contexts within which they functioned. They were imagined and experienced in unique ways by those designing them and those they targeted. There are no straightforward conclusions here. Industrial education was an extractive feature of national citizen-making and imperial subject-making, but its variants were also adopted by anticolonial movements in Asia, Black activists in the United States and colonized Africa, and Latin American politicians and educators seeking economic parity with the North Atlantic empires. Nevertheless, the many iterations detailed in this introduction and in the wider volume not only shared a familial resemblance, but also directly dialogued with and drew from one another. Understanding these modes of industrial education in comparative and connective contexts is crucial.

The ultimate durability of industrial education arose from its survival instincts and adaptability. The mid-twentieth-century abeyance of the colonial empires gave way to Cold War models of development and, ultimately, a present-day capitalized modernity that relies on forms of training to facilitate the unequal labor mobilities underwriting corporate and national power. *The Gospel of Work and Money* is the first collection of its kind to consider these trajectories of industrial education in the modern world. It suggests the need for new scholarly conversations about historical and contemporary iterations of the practice, which are so numerous as to resist containment in a single volume. We hope this book encourages them.

Notes

1. W. E. B. Du Bois, *The Souls of Black Folk* (Oxford University Press, 2007), 39. Emphasis ours.

2. Du Bois's explanation of the "Tuskegee machine" can be found in his autobiography, *Dusk of Dawn: An Essay Toward an Autobiography of a Race Concept* (Oxford University Press, 2007), 36 (originally published in 1940); quote taken from Du Bois, *The Souls of Black Folk*, 39.

3. For Washington's response to Du Bois's critiques, see Kevern J. Verney, *The Art of the Possible: Booker T. Washington and Black Leadership in the United States* (New York: Routledge, 2001), 79–94. Du Bois later recognized how these differences shaped their conflicting philosophies as well. See Ralph McGill, "W. E. B. Du Bois," *Atlantic Monthly* 216 (November 1965): 79.

4. For Washington's justifications of industrial schooling, see his "Atlanta Compromise" speech, reproduced with his context in Booker T. Washington, *An Autobiography: The Story of My Life and Work* (Atlanta: J. L. Nichols & Co., 1900), 153–186.

5. Booker T. Washington, "Industrial Education for the Negro," in *The Negro Problem: A Series of Articles by Representative American Negroes of To-Day* (New York: James Pott & Company, 1903), 17.

6. Marcus Garvey, *Philosophy and Opinions of Marcus Garvey, or Africa for the Africans*, compiled by Amy Jacques Garvey (London: Frank Cass and Company, 1967), 54; Wilson Jeremiah Moses, *Creative Conflict in African American Thought* (Cambridge: Cambridge University Press, 2004), 260.

7. Du Bois, *The Souls of Black Folk*, 66–67. For scholarly work on racial capitalism, see Eric Williams, *Capitalism and Slavery* (Chapel Hill: University of North Carolina Press, 1994); Destin Jenkins and Justin LeRoy, eds., *Histories of Racial Capitalism* (New York: Columbia University Press, 2021). On the links between racial capitalism and education in the United States, see William H. Watkins, *The White Architects of Black Education: Ideology and Power in America, 1865–1954* (New York: Teachers College Press, 2001).

8. Quoted in Lewis Flint Anderson, *History of Manual and Industrial School Education* (New York: D. Appleton and Company, 1926), 21–22.

9. Anderson, *History of Manual and Industrial School Education*, 28.

10. Anderson, *History of Manual and Industrial School Education*, 28–32.

11. Kalani Beyer, "Setting the Record Straight: Education of the Mind and Hands Existed in the United States before the 1880s," *American Educational History Journal* 37, no. 1 (2010): 150.

12. Wolf-Dietrich Greinert, *Mass Vocational Education and Training in Europe: Classical Models of the 19th Century and Training in England, France and Germany During the First Half of the 20th* (Luxembourg: Office for Official Publications of the European Communities, 2005), 9–21; James Foreman-Peck, "Spontaneous Disorder? A Very Short History of British Vocational Education and Training, 1563–1973," *Policy Futures in Education* 2, no. 1 (2004): 79–82.

13. "Industry," Cambridge Dictionary Online, https://dictionary.cambridge.org/dictionary/english/industry.

14. Anderson, *History of Manual and Industrial School Education*; Charles A. Bennett, *History of Manual and Industrial Education up to 1870* (Peoria, IL: Chas. A. Bennett Col, Inc., 1926); Charles A. Bennett, *History of Manual and Industrial Education 1870–1917* (Peoria, IL: The Manual Art Press, 1937). On this trend in U.S. historiography see Oli Charbonneau, "Teaching the World to Work: Industrial Education as U.S. Imperial Tradition," *Diplomatic History* 47, no. 3 (2023): 369–390.

15. Clayton Pierce, "W.E.B. Du Bois and Caste Education: Racial Capitalist Schooling from Reconstruction to Jim Crow," *American Education Research Journal* 54, no. 1 (April 2017): 27S.

16. Naoko Shibusawa, "U.S. Empire and Racial Capitalist Modernity," *Diplomatic History* 45, no. 5 (2021): 863–884.

17. In sketching a global history of industrial education and its legacies, our volume follows work like Barnita Bagchi et al., eds., *Connecting Histories of Education: Transnational and Cross-Cultural Exchanges in (Post)Colonial Education* (New York: Berghahn Books, 2014).

18. Stephanie Olsen, "Introduction," in *Childhood, Youth and Emotions in Modern History: National, Colonial and Global Perspectives*, ed. Stephanie Olsen (London: Palgrave Macmillan, 2015), 4–6; Karen Vallgårda, Kristine Alexander, and Stephanie Olsen, "Emotions and the Global Politics of Childhood," in *Childhood, Youth and Emotions in Modern History: National, Colonial and Global Perspectives*, ed. Stephanie Olsen (London: Palgrave Macmillan, 2015), 13–16.

19. Hugh D. Hindman, *Child Labor: An American History* (Abingdon-on-Thames: Routledge, 2002), 12–35; James D. Schmidt, *Industrial Violence and the Legal Origins of Child Labor* (Cambridge: Cambridge University Press, 2012), 4–31; on the early history of industrialized child labor see: Kristen McCabe Lashua, *Children at the Birth of Empire: British Law, Liberty, and the Global Migration of Destitute Children, c. 1607–1760* (Abingdon-on-Thames: Routledge, 2023).

20. Peter N. Stearns, *Childhood in World History*, 4th edition (Abingdon-on-Thames: Routledge, 2021), 106–107.

21. See for example Sinéad Pembroke, "Foucault and Industrial Schools in Ireland: Subtly Disciplining or Dominating Through Brutality?" *Sociology* 53, no. 2 (2019): 385–400; William McGovern, "On Distant Shores: The Transatlantic Foundations of Child Reform in the Mid-Nineteenth Century," *Journal of Transatlantic Studies* 17, no. 4 (2019): 506–531; Andrew G. Ralston, *Opening Schools and Closing Prisons: Caring for Destitute and Delinquent Children in Scotland, 1812–72* (Abingdon-on-Thames: Routledge, 2017); Cameron Nunn, "Juvenile as Human Capital: Re-evaluating the Economic Value of Juvenile Male Convict Labour," *Labour History* 108 (2015): 53–69.

22. Mary Nejedly, *The Industrious Child Work: Child Labour and Childhood in Birmingham and the West Midlands, 1750–1900* (Hertfordshire: University of Hertfordshire Press, 2021), 55–88; Beate Althammer, "Introduction: Poverty and Deviance in the Era of the Emerging

Welfare State," in *The Welfare and the 'Deviant Poor' in Europe, 1870–1933*, ed. Beate Althammer et al. (Basingstoke: Palgrave Macmillan, 2014), 2–9.

23. Elizabeth C. Brown, "Middle Passages: Lessons in Racial Subjection at the Hampton Institute and Carlisle Indian Industrial School," *American Quarterly* 75, no. 4 (2023): 724–725.

24. More broadly, colonial education is also inextricably linked to colonial order making and exploitation. See Ana Isabel Madeira and Luís Grosso Correia, "Colonial Education and Anticolonial Struggles," in *The Oxford Handbook of the History of Education*, ed. John L Rury and Eileen H. Tamura (Oxford: Oxford University Press, 2019), 413–426; Nicholas Harrison, *Our Civilizing Mission: The Lessons of Colonial Education* (Liverpool: Liverpool University Press, 2019), 285–319.

25. Sjaak Braster, "The People, the Poor, and the Oppressed: The Concept of Popular Education Through Time," *Pedagogica Historica* 47, no. 1/2 (2011): 10.

26. Kirsten Kamphuis, "Giving for Girls: Reconsidering Colonial Civilizing Missions in the Dutch East Indies Through Charitable Girls' Education," *New Global Studies* 12, no. 2 (2018): 219.

27. See Antoinette Burton, ed., *Gender, Sexuality, and Colonial Modernities* (Abingdon-on-Thames: Routledge, 1999); Michele Mitchell and Naoko Shibusawa et al., "Introduction: Gender, Imperialism and Global Exchanges," in *Gender, Imperialism and Global Exchanges* (Hoboken: Blackwell Publishing, 2015), 1–21.

28. The idea of the "problem" is explored in Oli Charbonneau, "Colonizing Workers: Labor, Race, and U.S. Military Governance in the Southern Philippines," *Modern American History* 4, no. 1 (2021): 25–47.

29. For reflections on the rise of racial capitalism under conditions of empire see Susan Koshy et al., *Colonial Racial Capitalism* (Durham: Duke University Press, 2022).

30. Felicity Jensz, "Missionaries and Indigenous Education in the 19th Century British Empire: Part I: Church-State Relations and Indigenous Actions and Reactions," *History Compass* 10, no. 4 (2012): 311.

31. While apprentice training dates to the Middle Ages, experiential training as part of school curricula originated in the early modern period. In Halle, Germany, the clergyman August Hermann Francke established his Paedagogium in 1715, which included a "shop for the cutting and polishing of glass" by students. See Gabriel Compayré, *The History of Pedagogy* (Boston: D.C. Heath & Company, 1889), 414.

32. Beyer, "Setting the Record Straight," 150–152. Pestalozzi's ideas spread well beyond German-speaking areas, inflecting later educational developments in Britain and the United States, among other places. See J. A. Brown, "British Pestalozzianism in the Nineteenth Century: Pestalozzi and His Influence on British Education" (PhD diss., University of Wales, 1986), and Karl-Heinz Füssl, "Pestalozzi in Dewey's Realm? Bauhaus Master Josef Albers Among the German-Speaking Emigrés' Colony at Black Mountain College (1933–1949)," *Paedagogica Historica* 42, no. 1–2 (2006): 77–92.

33. The idea of "tools as the great civilizing agency of the world" was common in nineteenth-century European and Euro-American reform movements, as typified in Charles H. Ham, *Manual Training: The Solution of Social and Industrial Problems* (London: Black & Son, 1886). In the eyes of Ham and his ilk, educating the laboring classes neatly resolved the "problems" created by industrial modernity.

34. Wolf-Dietrich Greinert, *Mass Vocational Education and Training in Europe: Classical Models of the 19th Century and Training in England, France and Germany During the First Half of the 20th*, 9–15.

35. Foreman-Peck, "Spontaneous Disorder?," 78.

36. Jillian Carol Gear, "Industrial Schools in England, 1857–1933: 'Moral Hospitals' or 'Oppressive Institutions'?" (PhD diss., University of London, 1999), 9.

37. Gear, "Industrial Schools in England," 10–11.

38. *Speeches of the Earl of Shaftesbury, K.G. Upon Subjects Having Relation Chiefly to the Claims and Interests of the Labouring Class* (London: Chapman and Hall, 1868), 191. The citation, and the broader historical context for the ragged school movement, found in Gertrude Himmelfarm, *The Idea of Poverty: England in the Early Industrial Age* (New York: Vintage Books, 1983), 376.

39. Christine Mayer, "Poverty, Education and Gender: Pedagogic Transformations in the Schools for the Poor (*Armenschulwesen*) in Hamburg, 1788–1871," *Paedagogica Historica* 47, no. 1–2 (2011): 91–95.

40. Tobias Karlsson, Fay Lundh Nilsson, and Anders Nilsson, "Vocational Education and Industrial Relations: Sweden, 1910–1975," *Nordic Journal of Educational History* 5, no. 1 (2018): 27–50.

41. Nicoleta Roman, "Labour After Education: Vocational Schools, Graduates and Economic Growth in Modern Romania (1878–1914)," Center for Economic Research and Graduate Education—Economics Institute, Charles University, www.cerge-ei.cz/pdf/gdn/rrc/RRC15_21_paper_01.pdf.

42. Anderson, *History of Manual and Industrial School Education*, 142–143.

43. Beyer, "Setting the Record Straight," 155.

44. Anderson, *History of Manual and Industrial School Education*, 155.

45. Steven Hahn, *A Nation Without Borders: The United States and Its World in an Age of Civil Wars* (New York: Penguin Books, 2016), 317–361.

46. Ana Isabel Madeira and Luís Grosso Correia, "Colonial Education and Anticolonial Struggle," in *The Oxford Handbook of the History of Education*, ed. John L. Rury and Eileen H. Tamura (Oxford: Oxford University Press, 2019), 414–417; Ewout Frankema, "Why Was the Dutch Legacy So Poor? Educational Development in the Netherlands Indies, 1871–1942," *Centre for Global Economic History Working Papers* no. 54 (2014), 1–2, www.cgeh.nl/sites/default/files/cgehwp54_frankema.pdf.

47. Madeira and Correia, "Colonial Education," 416–418.

48. Amanda Barry, "'Equal to Children of European Origin': Educability and the Civilising Mission in Early Colonial Austrialia," *History Australia* 5, no. 2 (2008): 414.

49. Jensz, "Missionaries and Indigenous Education," 294.

50. David Maxwell, "Missionary Movement in African and World History: Mission Sources and Religious Encounter," *Historical Journal* 58, no. 4 (2015): 901–930.

51. Jensz, "Missionaries and Indigenous Education," 296.

52. Ana Isabel Madeira, "Comparing Colonial Education Discourses in the French and Portuguese African Empires: An Essay on Hybridization," *Revisita Española de Educación Comparada* 31 (2018): 135.

53. Maaike Derkson, "'On Their Javanese Sprout We Need to Graft European Civilisation': Fashioning Local Intermediaries in the Dutch Catholic Mission, 1900–1942," *Tijdschrift voor Genderstudies* 19, no. 1 (2016): 25–48.

54. John Foreman, *The Philippine Islands: A Political, Geographical, Ethnographical, Social and Commercial History of the Philippines Archipelago* (New York: Charles Scribner's Sons,

1906), 194; Patricio N. Abinales and Donna J. Amorosa, *State and Society in the Philippines* (Lanham: Rowman & Littlefield Publishers, 2005), 92–95.

55. Eleazar Wheelock, *A Continuation of the Narrative of the Indian Charity School* (Hartford: E. Watson, 1775), 9–31; Keely McCarthy, "Conversion, Identity, and the Indian Missionary," *Early American Literature* 36, no. 3 (2001): 353–360.

56. Andrew K. Frank, "Indian Civilization Fund Act," in *Encyclopedia of American Indian History*, ed. Bruce E. John and Barry M. Pritzker (Oxford: ABC-CLIO, 2008), 540–541; Christina Snyder, *Great Crossings: Indians, Settlers, and Slaves in the Age of Jackson* (Oxford: Oxford University Press, 2017), 70–98.

57. Francis Paul Prucha, *The Great Father: The United States Government and the American Indians* (Lincoln: University of Nebraska Press, 1995), 482.

58. Department of the Interior, *Report of the Secretary of the Interior* (Washington, D.C.: October 31, 1873), 740.

59. John Eaton, *The Indian School at Carlisle Barracks* (Washington, D.C.: Government Printing Office, 1880), 3–5; Richard Henry Pratt, *Battlefield & Classroom: Four Decades with the American Indian, 1867–1904*, ed. Robert M. Utley (Norman: University of Oklahoma Press, 2003), 191–244.

60. Emily Conroy-Krutz, *Christian Imperialism: Converting the World in the Early American Republic* (Ithaca: Cornell University Press, 2015), 51–73.

61. Karine Walther, "'The Same Blood as We in America': Industrial Schooling and American Empire," in *Religion and US Empire*, ed. Sylvester Johnson and Tisa Wenger (New York: New York University Press, 2022), 151–152.

62. On this broad arc of racialized industrial education in the nineteenth-century United States see Oli Charbonneau, "Teaching the World to Work: Industrial Education as U.S. Imperial Tradition," *Diplomatic History* 47, no. 3 (2023): 375–382.

63. Katsuya Hirano, "Settler Colonialism in the Making of Japan's Hokkaidō," in *The Routledge Handbook of the History of Settler Colonialism*, ed. Edward Cavanagh and Lorenzo Veracini (New York: Routledge, 2017), 336.

64. Ann Curthoys and Jessie Mitchell, *Taking Liberty: Indigenous Rights and Settler Self-Government in Colonial Australia, 1830–1890* (Cambridge: Cambridge University Press, 2018), 180–207 and 361–385; C. Joseph Genetin-Pilawa, *Crooked Paths to Allotment: The Fight Over Federal Indian Policy After the Civil War* (Chapel Hill: University of North Carolina Press, 2012), 136–155; Darren Reid, "'Compound Dispossession' in Southern Ontario: Converging Trajectories of Colonial Dispossession and Inter-Indigenous Conflict, 1886–1900," *Journal of Canadian Studies* 57, no. 1 (2023): 81–113.

65. Rebecca Swartz, *Education and Empire: Children, Race and Humanitarianism in the British Settler Colonies, 1833–1880* (Cham: Palgrave Macmillan, 2019), 132.

66. "Canada's Residential Schools: The History, Part I: Origins to 1939," *The Final Report of the Truth and Reconciliation Commission of Canada*, Vol. 1 (Montreal & Kingston: McGill-Queen's University Press, 2015), 329–352.

67. Cynthia Cohen, "'The Natives Must First Become Good Workmen': Formal Education Provision in German South West and East Africa Compared," *Journal of African Studies* 19, no. 1 (1993): 130.

68. Andrew E. Barnes, "Christianity and Vocational Education in Africa," in *The Palgrave Handbook of African Education and Indigenous Knowledge*, ed. Jamaine M. Abidogun and

Toyin Falola (Switzerland: Springer International Publishing, 2020), 163. On this topic, see also Andrew E. Barnes, *Global Christianity and the Black Atlantic: Tuskegee, Colonialism, and the Shaping of African Industrial Education* (Waco: Baylor University Press, 2017).

69. Soni, "Learning to Labour: 'Native' Orphans in Colonial India, 1840s–1920s," *International Review for Social History* 65, no. 1 (2020): 34.

70. Felicity Jensz, "The 1910 Edinburgh World Missionary Conference and Comparative Colonial Education," *History of Education* 47, no. 3 (2018): 409.

71. Oli Charbonneau, "Logics of Immersion: Lake Mohonk and the U.S. Colonial Boarding School," in *Global Perspectives on Boarding Schools in the 19th and 20th Centuries*, ed. Felicity Jensz and Daniel Gerster (London: Palgrave, 2022), 213–235.

72. Cohen, "'The Natives Must First Become Good Workmen,'" 117.

73. Sybill Kuster, "'Book Learning' Versus 'Adapted Education': The Impact of Phelps-Stokeism on Colonial Education Systems in Central Africa in the Interwar Period," *Paedagogica Historica* 43, no. 1 (2007): 84; on American influence on British West Africa see: Shoko Yamada, "Educational Borrowing as Negotiation: Re-Examining the Influence of American Black Industrial Education on British Colonial Education in Africa," *Comparative Education* 44, no. 1 (2008): 21–37.

74. Quincy Wright, "The Government of Iraq," *American Political Science Review* 20, no. 4 (November 1926): 755.

75. Chitrlada Burapharat and Supat Chupradit, "Vocational and Cooperative Education in Thailand: A Presentation," in *Education, Economy and Identity: Ten Years of Educational Reform in Thailand*, ed. Supat Chupradit and Audrey Baron-Gutty (Bangkok: Institute de Recherche sur l'Asie de Sud-Est, 2008), 57–60.

76. Ethan W. Ris, "The Education of Andrew Carnegie: Strategic Philanthropy in American Higher Education, 1880–1919," *Journal of Higher Education* 88, no. 3 (2017): 403–408.

77. Andrew Barnes, "Thomas Jesse Jones, the Phelps Stokes Commissions, and Education for Social Welfare in Colonial Africa," *Oxford Research Encyclopaedia of African History* (2020), https://oxfordre.com/africanhistory/view/10.1093/acrefore/9780190277734.001.0001/acrefore-9780190277734-e-653.

78. Moon-Ho Jung, *Menace to Empire: Anticolonial Solidarities and the Transpacific Origins of the U.S. Security State* (Oakland: University of California Press, 2022), 7–25; Gabriella Treglia, "Cultural Pluralism or Cultural Imposition? Examining the Bureau of Indian Affairs' Education Reforms During the Indian New Deal (1933–1945)," *Journal of the Southwest* 61, no. 4 (2019): 821–862.

79. Sara Lorenzini, *Global Development: A Cold War History* (Princeton: Princeton University Press, 2019), 22–106; Daniel Immerwahr, *Thinking Small: The United States and the Lure of Community Development* (Cambridge: Harvard University Press, 2015), 1–15; Jeff Schuhrke, "'Comradely Brainwashing': International Development, Labor Education, and Industrial Relations in the Cold War," *Labor* 16, no. 3 (2019): 39–67.

80. Alyosha Goldstein, "On the Internal Border: Colonial Difference, the Cold War, and the Locations of 'Underdevelopment,'" *Comparative Studies in Society and History* 50, no. 1 (2008): 30.

81. See the June 2009 special issue of *Diplomatic History* on global histories of modernization, particularly Brad Simpson, "Indonesia's 'Accelerated Modernization' and the Global Discourse of Development, 1960–1975," *Diplomatic History* 33, no. 3 (2009): 467–486, and Daniel

Speich, "The Kenyan Style of 'African Socialism': Development Knowledge Claims and the Explanatory Limits of the Cold War," *Diplomatic History* 33, no. 3 (2009): 449–466.

82. Jamie Kizzrie, "A Look Back: SPLC Case Provided Homes for Hundreds of African-American Children in Alabama," Southern Poverty Law Center, October 27, 2013, www.splcenter.org/news/2013/10/27/look-back-splc-case-provided-homes-hundreds-african-american-children-alabama; "Mt. Meigs Campus," Alabama Department of Youth Services, 2023, https://dys.alabama.gov/mt-meigs-campus.

83. Jon Reyhner and Jeanne Eder, *American Indian Education: A History*, 2nd edition (Norman: University of Oklahoma Press, 2017), 223–267. In October 2024, President Joe Biden offered an apology for the oppression students faced at the boarding schools. In December 2024, he designated Carlisle Indian Industrial School a national monument to acknowledge its role in producing communal trauma. See "Biden Designates National Monument at Former Indian Boarding School," *Washington Post*, December 9, 2024, www.washingtonpost.com/politics/2024/12/09/biden-designate-national-monument-former-indian-boarding-school.

84. *Canada's Residential Schools: The History, Part 2, 1939 to 2000*; *The Final Report of the Truth and Reconciliation Commission of Canada* (Montreal: McGill-Queen's University Press, 2015), 9–106; Anna Haebich, "Forgetting Indigenous Histories: Cases from the History of Australia's Stolen Generations," *Journal of Social History* 44, no. 4 (2011): 1033–1046.

85. Young-Sun Hong, *Cold War Germany, the Third World, and the Global Humanitarian Regime* (Cambridge: Cambridge University Press, 2015), 11–48; Lorenzini, *Global Development*, 33–49.

86. Christina E. Crawford, *Spatial Revolution: Architecture and Planning in the Early Soviet Union* (Ithaca: Cornell University Press, 2022), 121–216; Katherine Lebow, *Unfinished Utopia: Nowa Huta, Stalinism, and Polish Society, 1949–56* (Ithaca: Cornell University Press, 2013), 1–12.

87. Targeting youth from marginalized populations for reeducation at boarding schools remains a practice in the PRC. See Chris Buckley, "How China Is Erasing Tibetan Culture, One Child at a Time," *New York Times*, January 9, 2025, www.nytimes.com/interactive/2025/01/09/world/asia/tibet-china-boarding-schools.html.

88. Raúl Delgado Wise, "Imperialism, Unequal Exchange, and Labour Export," in *The Oxford Handbook of Economic Imperialism*, ed. Zak Cope and Immanuel Ness (Oxford: Oxford University Press, 2022), 251–266; Christoph Scherrer, "Surplus Labour: Imperialist Legacies and Post-Imperialist Practices," in *The Oxford Handbook of Economic Imperialism*, ed. Zak Cope and Immanuel Ness (Oxford: Oxford University Press, 2022), 267–280.

89. *Twelfth Annual Report of the Poor Law Commissioners* (London: W. Clowes and Sons, 1846), 1–43.

PART I

Models

CHAPTER 1

The Industrial School at Sea: Empire, Poverty, and the Experiences of Boys on Training Ships

Christine Whyte

Introduction

In the second half of the nineteenth century, Industrial and Reformatory School Training Ships were launched across Britain as part of the broader implementation of industrial education as a seeming panacea for juvenile poverty, crime, and idleness. For the people who established industrial schools in Britain, these three things were inextricably linked, and, at a time when the British Empire was rapidly expanding, colonial and imperial concerns were at the forefront of their minds. Industrial education was, as the introduction to this volume makes clear, about more than just training for work. Its proponents in government, in religious and philanthropic institutions, and in the education system came to regard labor as having transformative potential. The potential to assimilate out-groups into a dominant culture; to reform deviant behavior; or to make impoverished communities self-sufficient. In this context, training ships were launched to prepare children to expand the reach of the empire's naval forces, but they were also intended to reform children seen as at risk of criminal behavior; to remove them from the presumed toxic influences of poor and working-class urban communities; and to set them on a different life path. But why specifically were ships outfitted to provide industrial education? How did philanthropists land on schemes to train children for naval service? How did the boys on these ships respond to being forcibly confined as a result? Tracing the launch of the training ships reveals some of the imperial imperatives that underpinned industrial education in Britain.

Sources about industrial school and reformatory training ships consist mainly of newspaper accounts, administrative files such as registers and minute books, and parliamentary papers and reports, all of which reveal little about the ships from the children's perspective. For example, when a parliamentary inquiry into Reformatory and Industrial Schools, including training ships, was launched in 1895, none of the 105 witnesses called were children who lived in one of the schools. In contrast, five Superintendents of Industrial School Training Ships were interviewed, and one Medical Officer.[1] This "silence" is telling. While industrial education was ostensibly for the benefit of poor and destitute youth, there was little appetite to find out what these children really thought of it. In an attempt to balance this discrepancy, this chapter draws on records that offer some partial insight into the perspectives of the boys.[2] These include records of an investigation into abuse on the Training Ships (TS) *Arethusa* and *Cornwall* and an unpublished manuscript written by H. A. Martin, who spent four years on TS *Mars*, moored in the River Tay, close to Dundee in Scotland, leading up to its closure and destruction in 1929.[3] This typescript account of his experience of the ship provides individual and specific detail from the perspective of a "Mars boy," as these children came to be known. In addition, we can draw our attention toward the unwritten ways that boys expressed their feelings about the ships, such as acts of vandalism, arson, and escape.

Reflecting on the experiences of children challenges prevailing views of the ships. As Siân Pooley and Jonathan Taylor argue, "we think differently about the history of modern Britain when we place children at the centre of our narratives."[4] Existing historical assessments of industrial schools and reformatory training ships have generally taken an official perspective. For example, in a general historical survey of nautical training ships, Phil Carradice concluded that industrial school ships were "an attempt to provide boys at risk with an acceptable and rewarding way out of the poverty trap."[5] However, references throughout the book point to a very different conclusion. Carradice himself described the regime on the TS *Mercury* as "perilously close to sadism."[6] In a similar survey of the certified industrial training ships, I. D. Cowan argued that "these ships appear to have successfully achieved their original aims—to remove boys who would otherwise have certainly become criminals from their environment."[7] These assessments of the training ships illustrate the "civilizing" aim of industrial education, but also highlight how later authors came to assess their "success" or otherwise by these official criteria. This chapter shifts focus onto how children responded to the goals of the training ships, the ways they subverted the founders' intentions, and how they interacted with each other.

Accounts of the children's experiences reveal the carceral nature of industrial education regimes. "Blurred lines" between punishment and education permeated industrial education in the British Empire.[8] Cowan noted that the Education Act of 1876 "allowed School Boards to take over the power of the prison authorities" and that the training ships in particular operated under a "tripartite area of responsibility," which made it hard to establish who was responsible for the running of the ships. Further, he draws attention to the continuing difficulties caused by the entanglements between education and prison authorities at the time he was writing in the 1980s.[9] These problems persisted, and a 2018 Who Cares? Scotland report and a Howard League briefing confirm that the criminalization of Care Experienced children and young people continues to be a pressing issue.[10] The training ships were part of a historical trend in which industrial education became associated with the punishment of criminalized children.

Several individual ships have drawn the attention of local historians, who have compiled detailed accounts of the ships' tenures, such as Emrys Wyn Roberts's study of TS *Clio* in Wales.[11] On Dundee's TS *Mars*, where Martin lived for four years, Gordon Douglas has written two well-illustrated popular history books drawn from the Lamb Collection and the Mars Collection in Dundee and Newport-on-Tay.[12] Douglas used these collections of newspaper clippings, photographs, minute books, and registers to illustrate the running of the ship. The title of one of the books, *We'll Send Ye Tae the Mars: The Story of Dundee's Legendary Training Ship*, recalls a popular threat parents used to keep unruly children in line. Local folk associations of the training ships with criminal or "bad" boys continue to the present day. The records also allude to the cruelty endured by the children, though Douglas contends that they were "in a better place" confined to the ship than living on the streets.[13] This assessment has a broader resonance. A 2023 policy briefing on UK government plans to use an "accommodation barge" to house asylum seekers pointed out that this twenty-first century policy was linked to a longer historical trajectory of turning "to ships to ease housing crises for so-called 'problem' groups."[14] Training ships were part of this deep-rooted trend and so deserve closer examination. This chapter takes an original approach to industrial education by positioning "industrial schools at sea" as a product of imperial ideology, which helps to explain why initiatives that appeared philanthropic in conception became carceral when put into practice.

The policy of confining poor, destitute, and criminally convicted children to residential training ships emerged from three interconnected trends in Britain: imperial, carceral, and philanthropic. The first section of this study, "An Empire of Poor Children," traces the imperial imperatives that converged to launch the training ship system: the transportation of criminalized children to far-flung colonies; the impact of the changing contours of the empire on naval and military recruitment; and the development of the ideology of forced training for work in the aftermath of the abolition of slavery. The second section, "Life Onboard the Training Ships," draws on children's testimony, as well as official documentation, to show how boys experienced the daily routines of the training ship as imprisonment and incarceration. These accounts show how far the training ships came adrift from the philanthropic ideals of the social reformers who first conceived of the industrial school system. The final section, "Beating Against the Wind," explores critiques of the training ship, both official and unofficial, that brought the training ship system to an end.

An Empire of Poor Children

British imperial ambitions and the fate of poor children were intertwined. Wherever empire expanded—from the North American colonies in the seventeenth century to Southern Africa in the early twentieth century—British children and young people were forcibly transported and labored in agriculture, domestic service, and onboard merchant ships.[15] Some of this forced child migration had a veneer of philanthropic purpose, offering an alternative to homelessness or destitution in Britain.[16] However, the impulse to remove poor and vagrant children mirrored the drive to remove "idle vagabondes" and the criminally convicted from Britain to face sentences of hard labor in the colonies from the sixteenth century onward.[17] Among these forced migrants were growing numbers of children.[18] An estimated ten percent of the female convicts transported to Australia in the early nineteenth century were girls under the age of eighteen, while between fifteen and twenty percent were boys.[19] These prisons for children provided an "industrial education" in manual skills, with the aim to produce useful white settlers for the colony. Children also frequently ended up imprisoned along with adults on convict "hulks," decommissioned ships anchored permanently near British cities. By the 1810s, there were calls to create specifically juvenile hulks, which were then launched in the 1820s.[20] But these schemes were regarded as reformative, not merely punitive. Cameron Nunn observes that the youngest of the first convict children sent to Australia, nine-year-old John Hudson, was sentenced to transportation by a magistrate who wished "to snatch such a boy, if one possibly could, from destruction."[21]

The criminalized poor were not only deployed as vanguard white settlers; the empire's military forces also used their labor. Hamish Maxwell-Stewart demonstrates that convict transportation was inversely related to military recruitment. At times of war, the British imperial state funneled poor and destitute people into the army and navy instead of penal colonies. Clare Anderson estimates that roughly fifteen thousand potential forced migrant convicts were redirected into military service between 1766 and 1826.[22] Alongside them, the naval and land forces also relied on the labor of enslaved and colonized people.[23] In the mid-nineteenth century, laws restricting their employment were repealed, and by 1886 there were an estimated sixteen thousand "lascars," or Indian mariners, working on British ships.[24] Africans were also a prominent feature of eighteenth-century British merchant shipping, and the British military was "greatly augmented" by the state's purchase of enslaved African men and boys in the late eighteenth and early nineteenth centuries.[25] After the British abolition of the slave trade in the Indian and Atlantic oceans, so-called "Liberated Africans" could also be impressed into naval and military service through a system of apprenticeship.[26] As noted in the introduction to this volume, apprenticeships and later "industrial education" were deployed at the end of slavery, amid claims that formerly enslaved people needed to be trained to work. These arrangements, which promised to fit people for freedom, simultaneously coerced them into manual labor.

Apprenticeship was also used for enlistment into the Royal Navy. From the early eighteenth century, the shipping industry was mandated to take on naval apprentices.[27] These apprenticeships bound boys and youths to merchant vessels for at least four years and provided trained recruits for the Royal Navy. But without an element of compulsion, this system started to fail. When the British engaged in the Crimean War in 1853, the shortfall in naval recruits became acute. Five years later, a Royal Commission recommended the creation of a system of school ships.[28] No longer able to rely on enslaved, forced, or impressed labor, the navy looked to the burgeoning industrial school and reformatory movement on land for a solution.

Sheriff William Watson introduced the industrial school system to Scotland in Aberdeen in 1841, but had envisioned a network of non-residential industrial day schools, located in urban centers and towns, that allowed children contact with family and community, alongside the food and training provided by the school.[29] In England, industrial schools took a different shape when the landed classes took up the cause, often constructing schools on their own estates. Much of the literature on British industrial education has paid scant regard to its international influences. However, much like the South American industrial-technological schools discussed by Helge Wendt in this volume, the British system of industrial education was shaped by inspiration from abroad. One key influence was the French reformatory Colonie de Mettray, opened in 1840, which stressed the reforming potential of rural life and, crucially, removal from family influence.[30]

Training ships were launched from some of the same imperatives as industrial schools. These could be broadly categorized into three types: reformatory schools for incarcerated children; industrial schools for the poor and working classes run by the government and by charities and voluntary organizations; and officer training schools and nautical colleges organized by the Royal Navy.[31] The first reformatory training ship, the *Akbar*, was launched two years after the 1854 Youthful Offenders Act and the Reformatory Schools (Scotland) Act introduced certified industrial and reformatory schools and placed them under the aegis of the Home Office.[32] Three reformatory training ships and sixteen industrial school training ships were launched between 1860 and 1885, many funded by an individual patron.[33] Another very similar pair of ships were the *Chichester* and the *Arethusa*, opened in Greenhithe, Kent, in 1866, managed by the Committee of the National Refuges for Homeless and Destitute Children (also known as "Shaftesbury Homes").[34]

Training ships were also deployed in the colonies; for example, the *Hastings* in Bombay (now Mumbai), the *Vernon* and *Sobraon* in Sydney, Australia, and the *Fitzjames* in Port Adelaide.[35] The *Akbar*, anchored in Liverpool, was built in Bombay and had served to escort prison ships to Australia,[36] and the *Mars* had been a supply ship in the Crimean War.[37] Many of the ships had seen service throughout the empire and were donated to voluntary organizations and charities for reuse as industrial schools.

Table 1.1: Reformatory and Industrial School Ships in the United Kingdom

Reformatory School Ships		
Ship name	*Location*	*Date founded*
Akbar, Liverpool Juvenile Reformatory Association (LJRA)	River Mersey, Liverpool	1856
Cornwall	River Thames, Essex	1859
Clarence, Liverpool Catholic Reformatory Association	River Mersey, Liverpool	1864
Industrial School Ships		
Ship name	*Location*	
Havannah	Cardiff, Wales	1860
Indefatigable	River Mersey, Liverpool	1865
Chichester and Arethusa I and II	River Thames, Greenhithe	1866 and 1873
Wellesley	River Tyne, South Shields	1868
Southampton	River Humber, Hull	1868
Cumberland/Empress	Helensburgh, Scotland	1869
Mars	Dundee, Scotland	1869
Formidable	Portishead, Bristol	1869
Goliath (destroyed by fire in 1875), Forest Gate School District	River Thames, London	1870
Gibraltar/Grampian	Belfast, Northern Ireland	1872
Clio	Bangor, Wales	1876
Warspite	Woolwich, Kent	1876
Exmouth (replaced Goliath), Metropolitan Asylums Board	Grays, Essex	1877
Mount Edgecumbe, Devon and Cornwall Industrial Training Ship Association	Saltash, Cornwall	1877
Shaftesbury, London School Board	River Thames, London	1878
Mercury	Binstead, Isle of Wight, then Southampton	1885

Life Onboard the Training Ships

To give an idea of what life on a training ship was like, we will take a look under the deck of the *Mars*, which housed more than six thousand boys between 1869 and 1929. The very first "Mars boy," Murdoch McLeod, was inspired by stories of imperial adventure to board the ship illegally in October 1869. The novel *Swiss Family Robinson* gave him the idea to "run away to sea."[38] Two years later, though, McLeod was involved in two incidents. First, he nearly drowned and was rescued from the water by another boy, Peter McKenna. McLeod then attempted to escape the boat by taking a life buoy and swimming through the night to Tayport, where he was found and returned by a police officer.[39] Much later in life, McLeod was interviewed about his time on the *Mars*, and recounted how he was "slung across cannon" and "lashed with a rope's end" as a boy.[40] At the time this interview was conducted, McLeod was in his eighties, remembering his life as a teenager. The details he recalled are striking: reading *Swiss Family Robinson*; the physical violence onboard; the "iron discipline" of the ship; and his unsuccessful attempts to get away.

Unlike the runaway McLeod, the first boy officially registered was categorized under Section 14 of the 1866 Industrial Schools Act as being destitute and homeless.[41] David Petrie had been living with his mother in Dron's Close, presumably in one of the "two or three dilapidated houses" in this "narrow opening."[42] His father's whereabouts were unknown. In theory, industrial schools were meant to house and educate the poor, as distinct from the reformatory schools for children convicted of crimes. In practice, children were sentenced to time in an industrial school by magistrates for everything from begging to being in the company of reputed thieves. This experience was very similar to a criminal proceeding. In 1874, twelve-year-old James Robertson pled "guilty" before an Edinburgh magistrate to not having a home or visible means of subsistence. He was sent to the *Mars* for a period of four years.[43] This case, like many others, shows that while the 1866 act was for destitute, not criminalized, children, the process was little different in practice. Magistrates would "charge" children with being vagrant, unhoused, or begging, and would "sentence" them to time in an industrial school, despite no criminal offense being committed. Local communities, and the boys themselves, understood it as punishment for bad behavior. In 1895, of the 127 boys committed to TS *Mars*, almost half were listed as "being found wandering"; forty-one had been "in the company of thieves," and sixteen were found begging. Only three were "destitute" or "orphaned," and only four had been charged with an offense.[44] At times, the process was curiously informal. In 1889, a boy named Stuart was committed to the *Mars* by Mr Briggs, a justice in Fifeshire, in a sentencing that took place in a private house rather than a court. Briggs knew almost nothing of the boy's circumstances, but "his friend at St. Andrews" had "nominated the boy."[45]

Other children were "referred" to training ships by philanthropic organizations. Inspectors from the Royal Scottish Society for the Prevention of Cruelty to Children (RSSPCC) sought out seemingly vagrant children, who they could take to magistrate's court for committal to industrial schools. This practice soon broke out into a scandal with accusations of the agents "snatching" children from the streets. Surveying the annual reports of the RSSPCC from this period reveals a particular bent to the organization's philanthropic work. Numerous entries refer to the "problem" of "tinkers," "gypsies," or "travellers," and inspectors rued their lack of legal power to remove the children of traveler folk and forcibly integrate them into settled life.[46] Kelly argues that the agents of the RSSPCC were "particularly invasive," inspecting people's homes, habits, and finances.[47] Their zeal resulted in large numbers of children committed to industrial schools. The 1908 report celebrated the introduction of the Children Act, noting that it would give the RSSPCC greater "power to send the children of tramps and vagrants in certain circumstances to Industrial Schools," rescuing them from "the contaminating influences of immoral surroundings."[48] Agents of philanthropic organizations were also empowered to send children to training ships.[49]

H. A. Martin came to the *Mars* from London. He was only eleven when he joined the ship, one of the youngest boys, though some ten-year-olds were also admitted despite the legal lower age limit of twelve. The 1894 report from the *Mars* revealed that of 386 children onboard, the ship housed forty-five boys under the age of twelve, and eighty-two between twelve and fourteen years old. They are described as "orphans—children of criminal or drunken parents, children deserted by their parents, illegitimate offsprings [sic] of unfortunate women, in fact children who only want to be taken from bad surroundings and given a fair chance to become decent members of society."[50] In Martin's case, his mother, who was separated from his father, met the ship captain and arranged for Martin to join the ship. At first he was excited by the prospect, but after taking the train alone from London to Dundee, he recalled that "my feelings were interspersed with a few tears."[51]

Upon arrival on the ship, the children were told to strip, after which a medical officer performed an inspection. H. A. Martin described the medical examination as a public event, with naked boys lined up waiting for the medical officer.[52] The results of the medical inspections revealed that the children in industrial schools were comparatively undernourished and stunted in their growth. Dr Branthwaite, medical advisor to the chief inspector of schools, argued in 1914 that "under improved conditions the effect of the bad start in life should be diminished or even nullified after two or three years of control under circumstances that should imply good food, some degree of warmth and comfort, regular but not excessive employment and exercise, and sufficient recreation."[53] But, far from being "improved" by life in an industrial school, Branthwaite reported that working-class children who were "free" enjoyed better physical health and development than those in the industrial schools and reformatories.[54] In response, the medical officer from the *Mars* claimed that the children's health was poor at intake, and that Branthwaite's analysis did not give this enough weight. Ultimately, the committee concurred with the *Mars* medical officer. They noted a "higher death rate" from disease, yet concluded that medical provisions were adequate and the cause must lie with the children's preexisting poor health.[55] It is difficult not to read this decision as influenced by the commonly held eugenics theory that physical wellbeing was directly tied to both hereditary factors and social environment.

A boy's individual identity was stripped away onboard the ship. In a practice very much parallel to prison, each child was assigned a number, which replaced their name. H. A. Martin wrote: "My given number was 58 and once that was given you lost all other identity, no more to be called by your Christian or surname—only 58."[56] From that point on, even just among the children themselves, they were referred to only by their assigned number.[57] A letter from Mr G. K. Goalen printed in the *Dundee Courier* described how the boys would call to each other by number while playing football, or on their annual trip to Elie, suggesting that even off the ship and in their leisure time, the children did not use names.[58] In Martin's account, he refers to himself as "58" on different occasions. This prison-like practice was not considered especially controversial. Much like the colonial Philippines case study examined by Sarah Steinbock-Pratt in this volume, the close resemblance of the industrial school at sea to the prison hulks of old was regarded as a feature, not a glitch.

Boys were deliberately given little opportunity to maintain connections with family and friends. Contemporary ideology inspired by the eugenics movement dictated that removing children from their home communities was essential for moral reform.[59] A. L. Scott, the *Mars* superintendent, testified that most of the boys on the ship were "children who only want to be taken from bad surroundings and given a fair chance to become decent members of society."[60] As Janne Lahti observes in this volume of the Carlisle residential school, industrial schools not only trained their students for employment, they also sought to disconnect students from their roots. Ships were quintessentially segregated spaces—removed from outside contact by expanses of cold water, the children could be housed conveniently close to cities without any risk of moral "contagion." An 1894 committee concluded that one of the primary advantages of the training ship was the "more thorough separation of the boys from the outer world."[61]

Onboard the ships, rebellious behavior, even mild insolence, was harshly punished. In the official register of punishments for 1873, six boys were given between seven and eighteen "stripes." While this type of physical abuse was common in schools and in the navy at the time, consideration should be given to the confined spaces in which these punishments were given. The ship offered few spaces for the children to be alone; all punishments would likely have been public affairs, meaning the children often saw a friend or companion whipped. A former staff member, Baillie Lewis, claimed that the official register did not reflect the true extent of violence on the ship and revealed a number of irregularities in how children were being assigned to the ship by magistrates.[62]

The children onboard industrial school training ships did not acquiesce entirely peacefully to this regime. The next section looks at a range of critiques of the industrial school at sea, from the official to the resistance of the boys onboard.

Beating Against the Wind

In 1884, TS *Clarence* went up in flames. Efforts to contain the blaze failed, as did attempts to scuttle and sink the ship to stop the fire.[63] Seven boys were convicted for the deliberate setting of the fire and removed to a disused reformatory school on land. A replacement ship had to be found. Fifteen years later, this ship too was destroyed by fire. Boys living onboard had been collecting old rags and flammable materials for some weeks, and set the ship alight to coincide with a visit from Bishop Allen of Shrewsbury. The *Clarence* fire was not unusual. A number of the reformatory and industrial school training ships burned, and the blame usually fell on the boys onboard.[64] This kind of overt sabotage of a ship was accompanied by a range of quotidian acts of disruption and conflict across all of the reformatory and industrial school training ships.

Boys refused to conform to the expectations of the industrial education regime. They stole food to supplement an inadequate diet, talked back to staff, and indulged in forbidden vices like smoking. They also made determined efforts to escape, whether for a day's respite or a longer period of time. Ship staff were often able to find and retrieve escapees, but they also relied on local police. In 1892, however, when Major Collier of the Shaftesbury Boys' Home wrote to a local police office and requested the arrest of one of the boys, the request was queried by Police Sergeant Webster because the boy had not been taken onto the ship by order of a magistrate.[65] A letter from TS *Mercury* likewise requested support from local police and was rebuffed: In this case, the police pointed out that they could take action against the escaped boy for "stealing clothing," but they had no power to arrest him, as the ship was an "entirely voluntary Institution."[66]

On several of the ships, "certain charges of immorality" were made against the boys onboard.[67] The sexual behavior of the boys was investigated and revealed that the boys formed their own emotional networks but also that sexual behavior was often actually widely prevalent in the form of bullying or abuse by staff members. Boys were afraid to report instances of sexual coercion or abuse. On the TS *Cornwall*, one boy reported to the police that he spent some time "trying to screw up [the] courage to tell the Captain" about the sexual abuse he suffered at the hands of the ship's schoolmaster.[68] Reluctance to discuss sex also further delayed action; Dr Norris of the *Cornwall* reported that "old-fashioned modesty" had prevented Captain Pyddoke from fully reporting the sexual nature of the teacher's crimes.[69] Matt Houlbrook further notes that one of the boys confined on the *Cornwall*, who later wrote memoirs of his life, did not mention the seemingly ubiquitous sexual behavior, consensual and nonconsensual, on the ship.[70] Reporting of abuse could be silenced in multiple ways: through fear, modesty, and shame.

Boys also railed against expectations by finding their way home after discharge. In August 1874, George Jack, secretary of the *Mars*, wrote that although 166 boys discharged from the ship had gone on to the Royal Navy or merchant service, fifty-seven (around a third) had taken up jobs on land. He blamed this on the influence of "depraved relatives."[71] An 1891 *Dundee Courier* newspaper article titled "The Children of Worthless Parents" stated that "authorities are thwarted by the action of the parents, who often choose to take the children to their own wretched homes."[72] The 1896 committee also raised this as an issue, pointing out that large numbers of those who did not go to sea "returned to friends."[73] Total separation from family and community was a hallmark of "success," but the training ships seem, even on their own terms, to have largely failed.

The training ship system was attacked in 1895 by Admiral Field, who pointed out that the ships were extravagant and wasteful, because so few of the boys eventually went to sea. For the *Mars*, around half of the graduates went to the sea, though the vast majority of those were in the Mercantile Marine, fishing, or associated business, not the Royal Navy.[74] By this time, the navy was in no need of new recruits, particularly not the undernourished and physically frail graduates of the industrial school at sea, and the navy would not accept recruits with a criminal record. Although being confined to an industrial school training ship was not meant to be a conviction, it was treated as such, and many boys who hoped to graduate into the navy were disappointed.

Field's critique was parried by Admiral Day Hort Bosanquet in 1898. He argued that "Great Britain is the maritime nation of the world . . . the calling of seamen is therefore one of the most important, if not the most important, of all careers open to her rescued boys."[75] His philanthropic vision was tempered with nationalist and racial anxieties.[76] Bosanquet quoted statistics that 27,765 foreigners were serving on British vessels, alongside 27,911 Lascars.[77] This, he argued, showed that there was a strong "demand for the services of boys in the Mercantile Marine" in order to replace both foreigners and British subjects of color in their ranks.[78] He suggested introducing more coercive powers of recruitment, for example, that the boys could make binding agreements without the cooperation of their parents or guardians. He recommended "packing" more boys onboard each ship, allocating just 7 feet by 25 inches for each boy's hammock space below decks. For his part, Bosanquet drew little distinction between the industrial school and the reformatory boy, saying that while the reformatory was for the "actual" delinquent, the industrial school was for the "potential delinquents."[79] As the training ships entered a long period of decline, proponents deployed the language of empire to defend them as bulwarks of "civilisation" against the "dissipating and distracting influences of modern life."[80] The imperial civilizing mission was brought home to the poor, white children of the training ships of Britain, and the coercive nature of industrial education at sea intensified.

At the beginning of the twentieth century, figures show a steep decline in the numbers of children admitted to industrial schools as a result of begging or vagrancy, but a sharp increase in those who were convicted of some kind of offense.[81] Martin described being asked, "what did you do?" when he arrived onboard, the assumption being that everyone must have committed some offense, even if it was just being destitute. At the same time, the treatment of children in the courts and reformatories was being questioned. The 1908 Children Act stipulated that children should go through an entirely separate court process, overseen by specialized and trained magistrates. The *Mars* executive committee meeting in 1909 noted that this immediately lessened the number of children being sent to the *Mars* by the School Boards.[82]

In October 1910, the newspaper *John Bull* reported that serious violent crimes against children had taken place on board TS *Akbar*. A full enquiry was ordered by First Lord of the Admiralty Winston Churchill. However, the allegations, including that a boy named Brown had died after being soaked in freezing water, were dismissed. Public protests ensued and forced Churchill to appoint a new committee to investigate the running of all industrial and reformatory schools, which led to the complete overhaul of the juvenile detention system along the lines for which Mary Carpenter had campaigned more than fifty years earlier.[83] This marked the beginning of the end for the training ships.

Conclusion

The labor needs of the British Empire shaped the way in which poverty in Britain was criminalized and work valorized as the "savior" of poor children. This can be seen in the continuation of emigration schemes for poor and destitute children, in the overlap between function and form of the industrial school and the reformatory, as well as in the creation and maintenance of a system of "training for empire," which used criminalized and poor children as vanguard settlers and enforcers of imperial power globally. One means of doing the work of empire was through incarceration, one way or another, of poor white children. This imperial imperative dovetailed more or less neatly with philanthropic ideology. Driven by a desire to save children from the perceived "evils" of urban environments, single-parent families, and impoverished communities, reformers ascribed to the new "gospel" of industrial education. Training ships were a very visible and literally outlandish feature of that trend. Tracing their origins demolishes the notion that imperial history is at odds with the history of the British working classes.

A focus on children's experiences gives a new perspective on imperial history. The ships show how naval authorities targeted poor British children to maintain the empire. Much of the discussion of the history of institutional care for children has focused on the unanswerable question of whether they are "better off" in care than on the streets. Some of the "sons of the *Mars*" and other training ships were keen to access new opportunities and adventure through the training ship. While many traveled far afield and did indeed staff the expansion of British power, others rejected that path and returned to their home communities, profoundly changed by their experiences at sea. The decision to train boys for naval service in the empire changed all of their lives, but at no point were they consulted about this "philanthropic" venture. Instead, much like emancipated people, or colonized populations, they were assumed to be grateful to receive "industrial education," even while that gift served imperial imperatives on home waters as well as abroad.

Notes

1. List of Witnesses in "Report to the Secretary of State for the Home Department of the Departmental Committee on Reformatory and Industrial Schools. Vol. II. Evidence and Index" (Eyre and Spottiswoode, 1896), iii.

2. It should be noted that local records offices and archives, as well as private collections of personal papers, contain letters from boys on training ships and reports which may contain more personal reflections. In addition to this, some ships have active groups of "alumni" and their families who discuss the history and experiences onboard the ships. These sources have potential to reveal more about children's experiences onboard individual ships.

3. The admissions book of the TS *Mars* lists a Henry Albert Martin, age 11 5/12, born on March 12, 1914, as having been admitted on October 7, 1925, sent by his mother as a "London Voluntary Case." His father's name is not listed, but his mother, Eva Martin, paid seven shillings and sixpence a week for his upkeep. He is recorded as No. 58. Mars Training Ship. Admissions & Discharges, 1918–1929, Newport Library and Heritage Centre.

4. Siân Pooley and Jonathan Taylor, eds. "Introduction," in *Children's Experiences of Welfare in Modern Britain* (London: University of London Press, 2021), 21.

5. Phil Carradice, *Nautical Training Ships: An Illustrated History* (Stroud: Amberley Publishing, 2009), 82.

6. Carradice, *Nautical Training Ships*, 97.

7. I. D. Cowan, "Certified Industrial Training Ships *c.* 1860–1913," *Journal of Educational Administration and History* 16, no. 1 (1984): 8.

8. Clarissa Carden, "Managing Moral Reformation: The Case of Queensland's Reformatory for Boys, 1871–1919," *History of Education Review* 50, no. 1 (2020): 226–240.

9. Cowan, "Certified Industrial Training Ships," 1.

10. Who Cares? Scotland (WCS), "Who Cares? Scotland's Report on the Criminalisation of Care Experienced People," April 26, 2018, www.whocaresscotland.org/wp-content/uploads/2022/06/WCS-Report-on-the-Criminalisation-of-Care-Experience-April-2018.pdf; and The Howard League for Penal Reform, "'This Is Our Story': Children and Young People on Criminalisation in Residential Care," 2018, https://howardleague.org/publications/this-is-our-story-children-and-young-people-on-criminalisation-in-residential-care.

11. Emrys Wyn Roberts, *The 'Clio' 1877–1920: A Study of the Functions of an Industrial Training Ship in North Wales* (Gwynedd: Llygad Gwalch Cyf, 2011); Bruce E. Hunt, *TS Mount Edgcumbe: Training Ship for Homeless and Destitute Boys* (Saltash: BEH Publications, 2007); Viki McDonnell, *Snatched from Satan: The Story of the Clyde Industrial Training Ship* (London: W. L. Publishing, 2023). Industrial education training ships were not confined to Britain; for example, the TS *Vernon* and TS *Sobraon* in Australia are examined in Sarah Luke's *Like a Wicked Noah's Ark: The Nautical School Ships Vernon & Sobraon* (Melbourne: Arcadia, 2020).

12. Gordon Douglas, *We'll Send Ye Tae the Mars: The Story of Dundee's Legendary Training Ship* (Edinburgh: Black & White Publishing, 2008); Gordon Douglas, *Sons of the Mars: Further Tales from the Tay's Legendary Training Ship* (Wormit: Shakey Hands Publishing, 2013).

13. Douglas, *Sons of the Mars*, 63.

14. Anna McKay, "Asylum Barges in Historical Context: Britain's Prison Hulks Expose Fault Lines in Today's Policy," August 7, 2023, *History & Policy*, www.historyandpolicy.org/policy-papers/papers/asylum-barges-in-historical-context-britains-prison-hulks-expose-fault-lines-in-todays-policy.

15. See, for example, Angela Austin, "Forgotten Children: Scotland's Colonial Child Servants, 1680–1760" (MPhil diss., University of Glasgow, 2017); Ellen Boucher, *Empire's Children: Child Emigration, Welfare, and the Decline of the British World, 1869–1967* (Cambridge: Cambridge University Press, 2014); Kristen McCabe Lashua, *Children at the Birth of Empire: British Law, Liberty, and the Global Migration of Destitute Children, c. 1607–1760* (Abingdon-on-Thames: Routledge, 2023); Gordon Lynch, *UK Child Migration to Australia, 1945–1970: A Study in Policy Failure* (Berlin: Springer Nature, 2021); and Joy Parr, *Labouring Children: British Immigrant Apprentices to Canada, 1869–1924* (Toronto: University of Toronto Press, 1994).

16. Peter Wilson Coldham, "The 'Spiriting' of London Children to Virginia: 1648–1685," *Virginia Magazine of History and Biography* 83, no. 3 (1975): 280–287; Michael Sparke, *The Poore Orphans Court, or Orphans Cry* (London: Printed by Anne Griffin for Michael Sparke junior, 1636); Gilliam Pugh, *London's Forgotten Children: Thomas Coram and the Foundling Hospital* (Cheltenham: History Press, 2011).

17. Richard Hakluyt, "A Discourse Concerning Western Planting, 1584," in *Collections of the Maine Historical Society*, ed. C. Deanne (Cambridge: John Wilson, 1877), 36–38, quoted in Hamish Maxwell-Stewart, "Convict Transportation from Britain and Ireland 1615–1870," *History Compass* 8, no. 11 (2010): 1221.

18. Kerin Gorton and John Ramsland point out that transportation was intended to civilize the convicted children, but also that children were seen as a potential civilizing force on the "pioneering frontier." See Kerin Gorton and John Ramsland, "Prison Playground? Child Convict Labour and Vocational Training in New South Wales, 1788–1840," *Journal of Educational Administration and History* 34, no. 1 (2002): 53.

19. Kim Humphery, "Objects of Compassion: Young Male Convicts in Van Diemen's Land, 1834–1850," *Australian Historical Studies* 25, no. 98 (1992): 13–33.

20. "Châtiment, réforme ou opportunisme? La délinquance juvénile et les pontons pénitentiaires anglais, 1822–1843" [Punishment, reform or opportunism? Juvenile delinquency and the English prison docks, 1822–1843], in *Catégoriser l'autre: aires anglophone et lusophone* [Categorizing the other: English-speaking and Portuguese-speaking areas], ed. Michel Prum (Paris: L'Harmattan, 2017), 129–162.

21. Cameron Nunn, "'Making Them Good and Useful': The Ideology of Juvenile Penal Reformation at Carters' Barracks and Point Puer," *History Australia* 14, no. 3 (2017): 329–343.

22. Clare Anderson, "Transnational Histories of Penal Transportation: Punishment, Labour and Governance in the British Imperial World, 1788–1939," *Australian Historical Studies* 47, no. 3 (2016): 381–397.

23. The government had attempted to restrict their numbers with the 1660 Navigation Act, which stipulated that three-quarters of the crew of British ships should be British. See April Lee Hatfield, *Atlantic Virginia: Intercolonial Relations in the Seventeenth Century* (Philadelphia: University of Pennsylvania Press, 2004), 50.

24. Gopalan Balachandran, *Globalizing Labour? Indian Seafarers and World Shipping, c.1870–1945* (New Delhi: Oxford University Press, 2012), 62.

25. Maria Alessandra Bollettino, "'Of Equal or of More Service': Black Soldiers and the British Empire in the Mid-Eighteenth-Century Caribbean," *Slavery & Abolition* 38, no. 3 (2017): 510–533; Charles R. Foy, "The Royal Navy's Employment of Black Mariners and Maritime Workers, 1754–1783," *International Journal of Maritime History* 28, no. 1 (2016): 31–35; David Lambert, *Soldiers of Uncertain Rank: The West India Regiments in British Imperial Culture* (Cambridge: Cambridge University Press, 2024), 1.

26. Richard Anderson, "The Diaspora of Sierra Leone's Liberated Africans: Enlistment, Forced Migration, and 'Liberation' at Freetown, 1808–1863," *African Economic History* 41 (2013): 105; Kyle Prochnow, "'Saving an Extraordinary Expense to the Nation': African Recruitment for the West India Regiments in the British Atlantic World," *Atlantic Studies* 18, no. 2 (2021): 149–171.

27. The recruitment of apprentices to the Navy was mandated in a series of laws from the early eighteenth century onward. V. C. Burton, "Apprenticeship Regulation and Maritime Labour in the Nineteenth Century British Merchant Marine," *International Journal of Maritime History* 1, no. 1 (1989): 29–49.

28. Nineteenth Century House of Commons Sessional Papers, "Report of the Commissioners Appointed to Inquire into the Best Means of Manning the Navy; Together with the Minutes of Evidence and Appendix," Command Papers, 2469, vol. 6, p. xv.

29. John A. Stack, "The Provision of Reformatory Schools, the Landed Class, and the Myth of the Superiority of Rural Life in Mid-Victorian England," *History of Education* 8, no. 1 (1979): 42.

30. Michel Foucault identified Mettray as the realization of the carceral system, and also as the source of the "principle of refamiliarization" in *Discipline and Punish: The Birth of the Prison* (London: Penguin, 2019). See also Stephen A. Toth, *Mettray: A History of France's Most Venerated Carceral Institution* (Ithaca: Cornell University Press, 2019).

31. In *Nautical Training Ships*, Carradice has separate chapters for the industrial school ships, reformatories, and what he calls charity and voluntary organization ships.

32. Andrew G. Ralston, "The Development of Reformatory and Industrial Schools in Scotland, 1832–1872," *Scottish Economic & Social History* 8, no. 1 (1988): 48; Kelly, "Criminalisation of Children in Scotland," 115.

33. Carradice, *Nautical Training Ships*; Julie Brumby, "Reformatory and Industrial Training Ships in Nineteenth-Century UK," March 31, 2021, https://ourcriminalancestors.org/2021/03/reformatory-and-industrial-training-ships-in-nineteenth-century-uk; Brumby's list does not contain TS *Indefatigable*, which is mentioned in Cowan, "Certified Industrial Training Ships *c.* 1860–1913."

34. See "Training Ships 'Chichester' and 'Arethusa,' Greenhithe/Upnor, Kent," www.childrenshomes.org.uk/TSArethusa. For some ships there is a considerable amount of archival material available in local archives. I have excluded from this list the various training ships associated with sea schools, such as HMS *Conway*, which were not used as reformatories or industrial school ships,

35. British Library, IOR/F/4/1935/83539, "Visit by the Governor of Bombay [Sir James Rivett Carnac] to the Indian Navy training ship Hastings on 11 Mar 1841"; Luke, *Like a Wicked Noah's Ark*; Nathan Richards and Sally K. May, "South Australia's 'Floating Coffin': The Diseased, the Destitute, and the Derelict 'Fitzjames' (1852–c.1900)," *Great Circle*, 25 no. 1 (2003): 20–39.

36. Charles Hardy, *A Register of Ships, Employed in the Service of the Honorable the United East India Company, from the Year 1760 to 1810* (London: W. Heseltine, 1811); Jorge Ortiz-Sotelo and Robert King, "'A Cruize to the Coasts of Peru and Chile': HM Ship *Cornwallis*, 1807," *Great Circle* 32, no. 1 (2010): 35–52.

37. See, for example, New York Public Library, *H.M.S. Mars and H.M.S. Formidable Logbook, 1862–1863, MssCol 1879*, Manuscripts and Archives Division, The New York Public Library.

38. Quotation taken from an interview with McLeod in Dundee in 1939, quoted in Douglas, *We'll Send Ye Tae the Mars*, 33. The reference to this 1812 novel (first translated into English in 1814) sets McLeod's ambitions in the context of the British Empire. The novel describes a London Missionary Society missionary and his family on their way to Australia when they are shipwrecked in the South Sea Islands. As Michelle Elleray argues, adventure novels played a key role in how empire was "conveyed to Victorian children." Michelle Elleray, *Victorian Coral Islands of Empire, Mission, and the Boys' Adventure Novel* (Abingdon-on-Thames: Routledge, 2019), 3.

39. Douglas, *We'll Send Ye Tae the Mars*, 62–63.

40. Douglas, *We'll Send Ye Tae the Mars*, 73.

41. The 1866 Industrial Schools Act extended the 1857 Industrial Schools Act from England to Scotland. These schools were intended for unhoused and destitute children, while children who had committed criminal offences were sent to the new "reformatories."

42. Scotland's Places, "OS1/14/35/10," https://scotlandsplaces.gov.uk/digital-volumes/ordnance-survey-name-books/forfarshire-angus-os-name-books-1857–1861/forfar-angus-volume-35/12.

43. Kelly, "Criminalisation of Children in Scotland," 115.

44. "Reformatory and Industrial Schools Committee. Report to the Secretary of State for the Home Department of the Departmental Committee on Reformatory and Industrial Schools. Vol. II Evidence and Index," Cmd 8290 (London: Her Majesty's Stationery Office, 1896), 211.

45. "Report to the Secretary of State for the Home Department of the Departmental Committee on Reformatory and Industrial Schools. Vol. II Evidence and Index," 677.

46. Glasgow Caledonian University (GCU) Archives, GB 1847 Royal Scottish Society for the Prevention of Cruelty to Children (RSSPCC), National Annual Reports 1885–1930.

47. Christine Kelly, "Continuity and Change in the History of Scottish Juvenile Justice," *Law, Crime and History* 6, no. 1 (2016): 72.

48. GCU Archives, RSSPCC National Annual Report, 1908, 8.

49. Kelly, *Juvenile Justice*, 154.

50. "Report to the Secretary of State for the Home Department," 15.

51. "H.A. Martin: Account of Life on the Training Ship 'The Mars'," St. Andrews Special Collections, MS 38385-3, Misc Papers, 5–6.

52. Martin, "Account," 7.

53. "Report of the Departmental Committee on Reformatory and Industrial Schools in Scotland, Cd. 7886" (London: His Majesty's Stationery Office, 1915), 64.

54. "Report of the Departmental Committee on Reformatory and Industrial Schools in Scotland, Cd. 7886," 49.

55. "Report of the Departmental Committee on Reformatory and Industrial Schools in Scotland, Cd. 7886," 50.

56. Martin, "Account," 9.

57. Gordon Douglas, "The Mars Training Ship and Elie," Elie and Earlsferry History Society, 2018, www.eliehistory.com/explore/mars-training-ship-elie-gordon-douglas.

58. Quoted in Douglas, *We'll Send Ye Tae the Mars*, 49.

59. Lynn Abrams, "'Blood Is Thicker Than Water': Family, Fantasy and Identity in the Lives of Scottish Foster Children," in *Child Welfare and Social Action in the Nineteenth and Twentieth Centuries: International Perspectives*, ed. Jon Lawrence and Pat Starkey (Liverpool: Liverpool University Press, 2001), 195–216.

60. "Report to the Secretary of State for the Home Department," 15.

61. Departmental Committee on Reformatory and Industrial Schools, *Report to the Secretary of State for the Home Department of the Departmental Committee on Reformatory and Industrial Schools Vol. 1* (London: Her Majesty's Stationery Office, 1896), 227.

62. Douglas, *We'll Send Ye Tae the Mars*, 66–68.

63. "A Reformatory Ship Destroyed by Fire," *Evening News* (London), January 18, 1884.

64. For example, see incidences of arson documented in Cowan, "Certified Industrial Training Ships," 4–5.

65. TNA, MEPO 2/294, "Absconders from the Arethusa Training Ship," Letter from Metropolitan Police, Caledonian Rd. Division, January 3, 1896.

66. TNA, MEPO 2/294, "Memorandum: Application by the Captain of the Voluntary Training Ship 'Mercury' for the arrest of Herbert Cumming," January 23, 1897. Ship staff also pointed out that they would "defray the expenses" of finding and arresting the boys.

67. See "The Training Ship Vernon," *Sydney Morning Herald*, April 16, 1869.

68. TNA, MH 102/161, Report of the Essex County Constabulary, June 18, 1923.

69. TNA, MH 102/161, "Personal Memorandum" from Dr. Norris.

70. Matt Houlbrook, "On the Emptiness of the Glory Hole, and Other Non-Problems," *Modernism/Modernity Print Plus*, 2016, https://modernismmodernity.org/forums/posts/emptiness-glory-hole-and-other-non-problems.

71. Douglas, *We'll Send Ye Tae the Mars*, 68.

72. "The Children of Worthless Parents," *Dundee Courier*, September 9, 1891, Newport Library and Heritage Centre.

73. "Report to the Secretary of State for the Home Department," 115.

74. Carradice, *Nautical Training Ships*.

75. Day H. Bosanquet, "Training Ships and the Mercantile Marine," *Charity Organisation Review* 3, no. 16 (April 1898): 177–207.

76. The argument that the training ships forestalled the navy being overwhelmed by "foreigners" was also deployed in Parliament in 1876; see Hansard, Vol. 227, March 10, 1876. https://hansard.parliament.uk/lords/1876-03-10/debates/0eea31f8-eb1e-4276-8213-ebff0b20cb61/TheMercantileMarine%E2%80%94TrainingShips%E2%80%94Observations.

77. Lascars were sailors from the Indian subcontinent, southeast Asia, Arab world, or British Somaliland.

78. Bosanquet, "Training Ships and the Mercantile Marine," 180.

79. Bosanquet, "Training Ships and the Mercantile Marine," 177–207.

80. "The Mars Is Doing National Work, says Dr G. F. Barbour, at Annual Meeting. The Problems Our Civilisation Has Brought," *Telegraph*, February 26, 1914, Newport Library and Heritage Centre.

81. Gillian Carol Gear, "Industrial Schools in England, 1857–1933: 'Moral Hospitals' or 'Oppressive Institutions'?" (PhD diss., University of London, 1999), 57.

82. "The School Board and the Mars. Not Sending the Boys," *Dundee Advertiser*, February 26, 1910, Newport Library and Heritage Centre.

83. Carradice, *Nautical Training Ships.*

CHAPTER 2

Working Toward Whiteness: Training and Outing in a Settler Colonial Industrial School in the United States

Janne Lahti

Introduction

"I have been work and try to earn some money since I came back from Carlisle." This is how Chiricahua Apache Janette Russell (nee Woods) responded in a questionnaire sent by her former place of education, the Carlisle Indian Industrial School in Pennsylvania. Writing from Oklahoma in 1910, fifteen years after her discharge from the school, she frequently returned to the question of work in her replies. Russell also lamented that she and her husband did not own the land, the cattle, or the buildings they worked on and lived in near Fort Sill, Oklahoma. This, her answers implied, was holding them back. Ramona Chihuahua, another former student of Carlisle and also a Chiricahua Apache, indicated much the same in her replies. She wrote of how she had "no land!, own some stock! no building! no money! all the money in the stock." Yet she too was a proud worker: "yes sir! Laundress, about 3 years." Another former student, Chiricahua Jason Betzinez, had become a blacksmith by using "what I learned at school the best I know how."[1]

Surveying the responses of the numerous Indigenous students who took similar questionnaires, one can detect how many former students drew intimate connections between industrial training, work, and respectable living. Proud of their work, and their willingness to work, some of these former students were disappointed that U.S. society did not live up to their expectations. Their training and work at Carlisle's industrial shops and outings had sold them ideas that linked industrial education with civilized living and status through work, where work was a springboard toward respectability and integration into settler society. In this sense, work at Carlisle was promoted as a path toward whiteness, where whiteness was not racially unmarked, but something potentially aspirational and reachable as a socially constructed colonial signifier of difference. In the late 1800s, whiteness in the United States was in flux. It was understood as a contested hierarchy of white ethnicities. As Mathew Frye Jacobson points out, whiteness was rethought throughout the United States when massive immigration of "undesirable" Eastern and Southern Europeans fractured all-inclusive formulations of Anglo-Saxon superiority and shifted the emphasis on degrees of difference among various white ethnicities. As whiteness became hierarchical, with Anglo-Saxons at the top, it also became more spacious, a more open category. At the same time, whiteness also emerged in the U.S. as a way for workers to make claims for equality to their employers, as first shown by David Roediger.[2] Thus, whiteness became a flexible notion closely associated with work, and, as such, something to reach for and aspire. Whites were not born, they were made—and this is what the gospel of work promulgated at Carlisle promised: a new life, an uplift from "savagery" to "civilization" in a world of empires.

It is a relatively common misperception to think that settler societies have not shown an interest in Indigenous labor. As noted in studies of the school, Carlisle assaulted Indigenous culture and identity by curtailing the use of native languages and renaming students with Anglicized names. It also immersed Indigenous children in Anglo-Saxon cultural rituals and bodily appearance, making them eat and dress like whites.[3] Yet, Carlisle was still at its core a training school. Its focus was vocational training and industrial skills. This was already evident in its official name, Carlisle Indian Industrial School. It was also apparent in the way government officials saw the school. Already shortly after the school's founding in 1879, the Commissioner of Indian Affairs, Hiram Price, made Carlisle's core mission clear in his yearly reports published in the Congressional Serial Set. He wrote at length of Carlisle as a "training school" doing "pioneering work" in industrial education.[4]

Carlisle was not unique, but representative of larger systems of settler colonial education and other colonial variants of industrial instruction targeting racialized minorities. By the early 1900s, there existed more than three hundred such institutions in the United States. Furthermore, these mainland schools influenced U.S. education systems overseas, including in the Philippines (see Steinbock-Pratt's chapter in this volume), where schools were also designed to manage, transform, and uplift the local nonwhite populace. Following on the path set by Carlisle, most of these schools promoted work as a civilizing force. Carlisle was eventually closed in 1918 and transferred back to the U.S. Army to be used as a hospital for soldiers returning from World War I. Yet, other schools long outlived it.[5]

This chapter takes a closer look at the work and training experiences of Indigenous students at Carlisle in the late 1800s and early 1900s. It discusses the on-campus workshops and off-campus outings, where students worked in white households, to show a Carlisle that was as much about learning how to work as about learning how to read. It exposes a Carlisle where industrial education and work were envisioned as a colonial civilizing process and a tool of settler colonial elimination, reflecting the intensively circulated ideas of cultural uplift among colonial empires at the time.[6] The Carlisle regime was built on white understandings of work, on the labor needs of the settler society, and on its gender-specific labor roles as Indigenous boys were trained in manual professions and farmwork and girls as domestics. The school did not cater to the needs of its Indigenous students; it did not ask what kind of work the students valued or what kind of skills they needed to benefit their own communities. Rather, it sought to disconnect the students from their roots and to fundamentally change them; to eliminate them as Natives and to rebirth them as clones of whites. The realities were often ripe for abuse and violence. When Carlisle closed, its cemetery had 196 students buried in it. Many others died during the outings.[7] Richard Henry Pratt, the first superintendent of the school, believed that Native students needed to be separated from their "savage surroundings" in tribal communities to be molded in the school toward respectable whiteness: God-fearing, family-oriented, and especially hardworking. Pratt and other school authorities believed that "what the Indian needed was to gain ability to hold his own, and fellowship with the whites."[8] In a sense, the students were supposed to work toward whiteness. Indeed, the adaptation of the work processes and the mastery of the skills taught at Carlisle ostensibly symbolized a transition toward whiteness from Indigenous backwardness.

Using primary sources ranging from personal memoirs to official reports, student files, school newspapers, and visual images, this chapter first discusses the meanings of Indigenous work in a settler colonial framing. It then describes work regimes and experiences at the Carlisle campus. Last, it examines Carlisle's outing program, an extension of these systems outside the school grounds that immersed Indigenous students in white households and workplaces as a temporary workforce.

Training and Work as Settler Colonial Elimination

Training at Carlisle represented the next step in U.S. settler colonial elimination of Indigenous peoples, a continuum from the warfare and physical violence that had taken place against Natives across the continent for generations. This elimination operated in the domain of the everyday rather than in the field of battle. Here, the settler colonial logic of elimination is less a zero-sum contest over the land than an erasure of Indigenousness, a contest over self-identities, of the ways people perceived themselves and were perceived by others. This kind of reading of Carlisle stresses how the matters of the everyday served as grounds for the creation and maintenance of colonial power and its racial identities.[9] In this context, whiteness can be seen as a performance and experience measured and made visible through body, behavior, and domesticity, but also through education and work.[10] School and work regimes—their creation and management—held a key role in determining racial and community membership and the manufacture and upholding of parameters between "us" and "them" across the colonial world.[11] As training and work furthered and transmitted certain ideas of what proper, respectable living should consist of, they developed into dense transfer points of power and anxiety, especially in places such as Carlisle that were built on uneven power relations, non-volunteer enrollment, and isolation of the children from their home communities.

The settler colonial paradigm represents the settlers as the driving force of historical change. According to the dominant narrative, before the settler there was no proper use of the land and no proper work. As settler colonialism revolves around Indigenous elimination and replacement, it relies on a notion that Indigenous peoples were passive and that historically settler societies have not depended on Indigenous labor. The foremost theorist of settler colonial studies, Patrick Wolfe, maintained that "their—Indigenous people's—historical role in settler colonialism is to disappear so far as the Europeans go, to get out of the way, to be eliminated, in order that the Europeans can bring their subordinated, coerced labor, mix that labor with the soil, which is to say set it to work on the expropriated land and produce a surplus profit for the colonizer."[12] For a long time, as Lawrence Kessler and Andrew Isenberg note, historians of the U.S. West argued that while Euro-Americans' desire for Natives' land was acute, settlers had no more than a passing interest in drawing on Natives as a source of labor.[13] And even if settlers relied on Native labor it was somehow exceptional. Many works on Native American history have countered these views, locating Indigenous people in a variety of worksites and professions in the settler economy, prior, during, and beyond the reservation period. And it is good to keep in mind that many of these Indigenous men and women worked in the settler economy without ever attending schools such as Carlisle.[14]

Yet, in the late 1800s, Anglo-Saxon America commonly believed that Indigenous peoples were destined to vanish, replaced by a white settler majority. However, many whites advocated saving the Native Americans from this oncoming extinction. In their vision, whiteness signified higher culture and universal norms, measured against which Native Americans were found lacking. Only through tutelage and guidance could Natives have a future—namely, through schooling and training that would make them more like whites. Native culture would die out in the process, but individuals would survive, changed and made anew, making a symbolic racial crossing from native "savagery" to white civilization. These racialized notions and the promise of individual inclusion in the white settler society, even if in its lower rungs, already drove the formulations of President Ulysses S. Grant's "Peace Policy" in the early 1870s. They gained more momentum with the Dawes Act of 1887, which created individual allotments for Indigenous households on communally held reservation lands. These measures sought the dissolution of Indigenous communities and transformation of Indigenous peoples into individual farmers and workers. They would take up traits of whiteness, like individual ownership of land, but they would also learn how to work like whites. In this, Carlisle sought to lead the way for future generations of Native Americans.

Federal authorities, such as Pratt, believed that childhood and youth presented an optimal time for making interventions into Indigenous people's lives. Much like with poor British children in Scotland (see Whyte's chapter in this volume), removing Native American children from their home communities was an essential component of their education and, through it, their reform toward whiteness.

These lines of thinking manifested in the birth of Carlisle Indian Industrial School in 1879, which quickly grew into the foremost industrial school for Indigenous youth in the U.S. empire. As an industrial school, Carlisle was meant to produce competent workers. Of course, there should be nothing curious about this, as, after all, this is what industrial schools do. But Carlisle did not aim to produce Native American laborers suited for the needs of their Native cultures and communities. Rather, it sought to fundamentally alter the students and hence the Native communities by producing a labor force reflecting the priorities, mores, and values of the settler society. Drawing from his personal experience in the U.S. Army fighting Indigenous peoples in the American West, Carlisle's founder and principal pedagogical motor, Richard Henry Pratt, personally saw work as a tool for elimination. In his view, it was through proper work that Native Americans would abandon their tribal affiliations and integrate into mainstream white society. For Pratt, the right kind of environment and especially the right kinds of tutelage and training played key roles. When explaining the Carlisle curriculum he had come up with a year after its founding in 1880, Pratt noted that besides teaching reading and writing English, arithmetic, and other academic subjects, the school's true focus was on vocational learning, on industrial skills: "It has been our object to give direction and encouragement to each student of sufficient age in some particular branch . . . of the mechanic arts."[15] As Jacqueline Fear-Segal notes, Pratt made Carlisle "a living experiment," embracing a dual mission. The school was intended to eradicate Indigenous tribal cultures and train Native students in white ways to equip them to participate in white society and ultimately to prepare them for potential U.S. citizenship in the future. Carlisle also served to demonstrate to white Americans that this transformation was both attainable and desirable to extinction.[16] Pratt wanted to show the Indigenous students how the white world worked, and to show the white world that the Indians could become part of this white world as good workers—that they had what it took to make it in the settler society.

Training Shops

"Our train stopped on a siding just outside the post of Carlisle Barracks . . . Then we marched through the gate . . . [and] assembled on the steps of the mess hall." This is how Jason Betzinez described his entry to Carlisle in April 1887. He also noted how soon after arrival the boys received "blue army-type uniforms" and the girls got "neat dresses" that they wore "proudly."[17]

As Figure 2.1 shows, students left their old clothes and belongings behind and entered a new material reality on the Carlisle campus. This was part of their path, and symbolic transition, toward whiteness. They were made to dress and eat like whites and they were organized into companies, drilled

Figure 2.1: Chiricahua Apaches four months after arriving at Carlisle [version 2], 1887. Source: Photographer John N. Choate. Carlisle Indian School Digital Resource Center, https://carlisleindian.dickinson.edu/images/chiricahua-apaches-four-months-after-arriving-carlisle-version-2-1887. Princeton University Library, Department of Special Collections, Box M28, Folder M0990.

and wearing polished uniforms. According to a school report, this "taught them obedience and cleanliness" while also giving "them a better carriage."[18] Students ate in large mess halls and were taught to sit quietly at the table and to use silverware. They attended classes in large buildings (see Figure 2.2) and slept in barracks, far removed from the brush shelters or teepees to which they had grown accustomed. This reorganization of their everyday lives consisted of vocational training, work clothing, work rules, and work tools, as well as rigid schedules and demands for obedience toward the instructors. According to the school catalogue, it was the industrial departments at the school that "teach the Indian youth to work. One-half of each school day is devoted to some branch of productive industry or effective labor," while students aged sixteen and older were directed to pursue a specific trade.[19]

Figure 2.2: Four workshop buildings at Carlisle, circa 1890. Photographer John N. Choate. Source: Carlisle Indian School Digital Resource Center, https://carlisleindian.dickinson.edu/images/four-workshop-buildings-c1890. Cumberland County Historical Society, CCHS_PA-CH2–015b.

Work was pursued in various industrial shops, occupying large buildings with interiors suited for learning specific skills (Figure 2.2). There was a blacksmith shop, carpenter shop, tailor shop, tinware/metal shop, and steam-heating plant, among others. The buildings were often renovated and new shops established. There were also fields on leased land near the campus for learning farmwork. Each shop had specifically determined practical and pedagogical tasks, as explained in the school catalogue from 1902. Take, for example, "Tinsmithing." Here, the catalogue states that "instruction covers the names of machines and tools, with their care and use." The students also learned of "the quality of tin; laying and making of patterns; cutting straight and curved lines." Thus, civilized work here was about mastering machinery and the ability to manufacture orderly produce. Relaying a sense of accomplishment, the catalogue also noted that "all the tinware used at the school is manufactured here." Not only did Carlisle trust the quality of work of its own students in this regard, but it also exported what they manufactured to other schools as well as Indian agencies on the reservations.[20]

The school catalogue provided detailed information on the kind of learning processes and work culture Carlisle wanted to teach its students. In shoemaking "much depends upon the accuracy of the eye. Attention is therefore given first to training the eye to shape the parts." After this is mastered by the pupil, "come repairing and the various steps in the manufacture of shoes progressing from the simpler to the more complex operations, including cutting and fitting."[21] Clearly, Carlisle wanted to give an impression that the everyday operation of these shops was systematically planned beforehand, and it proceeded with precision and logic. Instruction emphasized repetition and uniformity as the keys to industrial training, the ability to follow certain work processes mechanically and to come up with uniform results in series.

Under the guidance of white instructors, as shown in Figure 2.3 and Figure 2.4, the students worked on a plethora of tasks in the workshops, where these pedagogical patterns were put to practice, producing, for example, shoes, three different types of buckboards, and two models of wagon.[22] Thus, they learned important practical skills valued in white workplaces and society. Students also worked on varied construction details on campus, "attending to the large amount of repairing, building, and furnishing required at the school." This included the building used for the first workshops upon

Figure 2.3: Students making wheels outside the wagon shop at Carlisle, circa 1890. Source: Photographer John N. Choate. Carlisle Indian School Digital Resource Center, https://carlisleindian.dickinson.edu/images/students-making-wheels-front-blacksmith-and-wagon-shop-c1890. Cumberland County Historical Society, CCHS_PA-CH1-068b.

Figure 2.4: Students working in the shoe shop, circa 1881. Source: Photographer John N. Choate. Carlisle Indian School Digital Resource Center, https://carlisleindian.dickinson.edu/images/students-working-shoe-shop-c1881. National Anthropological Archives, Smithsonian Institution, NAA_73253; Photo Lot 81-12 06802900.

the school's opening.[23] While working in the shops, students could prove their worth and advance from "the grades of helper, apprentice, and efficient apprentice to that of journeyman."[24] So there was a hierarchy embedded in the workshop system, a ladder of progress that measured the acquired skill set and abilities of the students. In a sense, the system rated how the Indigenous youngsters advanced toward whiteness, with whiteness here indicating mastery of these work processes.

In the farming land adjacent to the campus, the students grew wheat, corn, oats, hay, and potatoes as the principal crops, with additional beets, cabbage, onions, sweet corn, and small fruits raised for the school kitchen. By 1902, the school also had two apple orchards, and between seventy and eighty hogs that supplied all the pork and lard needed for the school's subsistence. Boys who were not sent to outings worked in the fields for the summer, while other students had year-round training "in the handling of cows and in the care of milk and butter," with the school cattle herd reportedly numbering seventy-five heads.[25]

While boys trained to become farmers, blacksmiths, carpenters, and other skilled craftsmen, girls worked in the sewing room, which grew over the years into the domestic department. Here, the girls were taught to prepare "wholesome nutritious meals" and to arrange tables and wash dishes. Much like in the industrial shops, thoroughness and patterns were emphasized. This applied even in cooking, where the various techniques included "broiling, baking, boiling and frying." Although there was some room for variety, even this did not rely on individual whim but on following instructions. "Modifications, including mixing and seasoning" served to ensure "the cooking of food in a way most appetizing and digestible." The "course included the making of soups; cooking of meats, cereals and vegetables; the preparing of plain desserts; canning and preserving of fruit; methods of making doughs light," as well as whole menus for meals. There was also bread making, laundry work, nurse training, and, of course, sewing in the girls' training calendar.[26] This gender-specific training was meant to cater to the future of these girls in the domestic realm, as wives and mothers, but in practice also as domestic workers. It showed how gendered Carlisle's notions of whiteness were. There were no options for girls beyond the domestic work sphere.

School papers were the mouthpiece for the kind of respectability Pratt and the school leadership wished to promote. Take, for example, how *Schools News*, one Carlisle school paper, admiringly reported, "Our reporter took a walk around in the shops to see what the boys were doing. In all the shops every boy was busy. Not one of them looked like lazy boy, all looked industrious doing some thing." It was crucial to convey the close connection between whiteness and productivity. Here, the report individualized the workers by name, while nevertheless treating them as a uniform mass of obedient and busy hands. It stated that "In the carpenter shop there were Jock, (Arapahoe) Ralph, (Sioux) Elwood, (Iowa) and Joe Gunn, (Ponca) sawing out window and door frames. Oscar (Cheyenne) and Michael Burns (Apache) were busy carving balcony posts and Lester (Arapahoe) was out side chiseling out a beam. These things are all for our new hospital. The boys are pushing it up rapidly."[27]

No "lazy boys" at Carlisle seemed to be the school motto. Carlisle positioned its work routine in stark contrast to the prevalent stereotypes of Indigenous men being averse to and unsuited for demanding physical labor. These kinds of flawed perceptions that were very common among white Americans made assimilation synonymous with a white work ethic. Indigenous boys and girls were somehow naturally lazy, but, when they worked toward whiteness, this laziness was stamped out of them by industrial training.

In many ways, Carlisle was like a small industrial community, and the industrial output resulting from student work was quite impressive. For the fiscal year ending in March 1901, Carlisle reports state that the school had raised 688 goats, and produced 120 pairs of pantaloons, 1,504 pairs of shoes, 1,380 articles of tinware, 113 sets of harnesses, and 30 vehicles (wagons, buggies, and surreys) during the past twelve months.[28] "Made in Carlisle" did not, however, compete in the open market. Pratt did not wish to antagonize the surrounding white community in Pennsylvania by putting Carlisle products in open competition with local producers. Instead, he negotiated with the federal government to purchase items that could not be used at the school and to distribute them at Indian agencies across the country.[29] Thus, while the Indigenous children stayed away from their families for years at a time, the items they made at Carlisle actually reached their Indigenous communities in Western reservations.

The students' everyday work experience at Carlisle centered on the workshop buildings and the fields adjacent to the campus. But while mimicking the white work world, and while teaching its values and material cultures to the Indigenous students, the school remained an isolated environment, an insulated setting. It was not the real white world, although it sought to perform as such. For the real experience the students needed to be sent out, to be immersed in the white society and to live with white families. This would be essential for their industrial training and futures in the settler society.

Outings

Janette Russell arrived at Carlisle on November 6, 1886, at age fifteen. Her first outing assignment was ten months later to a white family in Torresdale, Pennsylvania. She did not come back to the school until December 1891, more than four years later. Four months onward, Janette was off again to another family. Of the nine years she was enrolled at Carlisle, Janette spent more than seven years on outing assignments with three different families in Pennsylvania. Thus, she spent the bulk of her time working away from school, living with a white family and immersed in white society, not in the classroom with her Apache or other Indigenous peers.[30]

The prevalence of outings in Russell's case was by no means exceptional, as the school's student records attest. While individual outing assignments typically lasted from spring to fall, it is not difficult to find outings in the official records that went on for more than a year at a time. A casual glance at the outing records from the 1890s shows, for example, that Victor Bear had one outing from September 1891 to September 1893 and another from March 1895 to June 1896, while Simon Beaufrey went out in September 1896 and returned to school in August 1898.[31] For her part, Ramona Chihuahua went on her first outing in May 1888, from which she returned to the school only in October 1891. Her subsequent outings kept her away from school for the better part of 1892, 1894, and 1895, before her discharge in November of the latter year, after nine years at the school.[32] School promotional literature notes that at peak times nearly half of the approximately eight hundred Carlisle students on outings remained out for the winter too.[33]

Outings were where Carlisle's ideas of Indigenous improvement toward whiteness through work were tested in the "real world." Pratt firmly believed that Native students would only benefit from an opportunity to live in white homes for an extended period of time in order to gain practical, firsthand experience in self-support and civilized living. Pratt noted that "Carlisle fills young Indians with the spirit of loyalty to the stars and stripes, and then moves them out into our communities to show by their conduct and ability that the Indian is no different from the white or the colored."[34] Pratt's slogan was: "To civilize the Indian, get him into civilization. To keep him civilized, let him stay."[35] Here, "civilization" equaled white families and communities. This, as Jacqueline Fear-Segal and Susan D. Rose have noted, "was the rationale for his 'Outing' program." In practice, nearly all Carlisle students went to outings and thus experienced multiple dislocations. Instead of returning home for the summers, they were sent "out" into local communities to work for white families.[36]

Ideally, the school propagated, the students in their "country homes" are "made members of the family and are carefully trained as sons and daughters of the family are trained."[37] Pratt preferred Quaker families as exemplifying proper virtues, meaning Christian goodness, family values, and hard work. By living in white households, the students would be told how to work and act white by their instructors, and simulate it in the school environment surrounded by hundreds of other Indigenous peoples when following the school schedule and rules. Outings put them into a different type of a situation, far away from campus. They were removed from any contact with other

Natives, and immersed in the sea of whiteness. If they stayed in an outing over the winter, students were expected to enroll in local public schools, making for an even more immersive experience.

The outing system proved very popular among white school administrators in the late 1800s. Already in 1881, the second year of the school's existence, Carlisle sent more than a hundred students to live with white families for the summer. By 1888, the number had increased to 447; in 1893 to 621; in 1900 to 893; and in 1903, the peak was reached with 948 pupils on outings.[38] The official Register of Pupils shows that during the 1890s, Carlisle had a whopping 5,092 individual outing assignments in total. The vast majority of those took the students to locales in Pennsylvania, but some also went to Maryland, New Jersey, and New York.[39] By the turn of the century at the latest, it was evident that Carlisle had become de facto a major provider of seasonal and temporary labor in the surrounding area. At first, Pratt preferred to place the Indigenous boys on farms so they could learn agricultural pursuits, the idea being that when they returned to their Indigenous communities in the West they could make a living from the soil and lead the farming revolution in their reservations by example. But over time, Pratt also pushed for more variation in the learning of different manual and industrial trades, following the pattern of training set up in the campus shops. Boys worked as machinists, wagonmakers, blacksmiths, electricians, and shoemakers, and even at steel mills. And in 1898, Pratt sent a dozen students to a summer resort to act as waiters and one to Philadelphia to work in a furniture factory. Soon thereafter, two Carlisle students worked for a man who trained them for the railroad service.[40]

Gaining plenty of publicity from 1915 to 1918, sixty-eight Carlisle boys did their outing at the Ford Motor Company in Detroit. Carlisle school papers made the most of this work detail, hailing it as a high point of Indigenous assimilation: "There probably never has been, in the history of the world, a democracy quite so purely and thoroughly democratic as the factory of the Ford Motor Company. In the Ford shop, making Ford cars, are titled Europeans, sons of American millionaires, refugees from the oppression and poverty of the Old World, American college students, and now American Indians."[41] The Carlisle Yearbook also identified some individual students "as a Ford man," calling it their "claim for distinction."[42]

The outing period at Ford also promised more permanent jobs at the factory for some of the students. This the school deemed "a most wonderful opportunity." The Yearbook professed that "many Indian boys are taking advantage of this great opportunity." It went further, propagating on the implications working at a place such as Ford could entail: "It means that we not only educate our students but, when they have demonstrated their fitness, we have a job waiting and place them in a position to go out and earn their living, competing with the white man on an even footing and with an equal chance of success."[43] Competing with the white man was exactly what Pratt had envisioned. Working toward whiteness meant not only learning the skills but working with the whites, competing with them in the job market.

While the school papers promoted the Ford experience, they advanced the merits of the outing system more broadly. A special "outing issue" of *The Carlisle Arrow and Red Man* in 1918 reads as promotional material for the transformative effects of the outing experience as a principal civilizer.[44] It provided students with a chance to do what was perceived as real white people's work in white workplaces, and sometimes, as in the Ford case, alongside white workers. Here and elsewhere outing was represented as privilege; it was not mandatory but something the students and the host families typically applied for. In theory at least, the outing system was under similar regulation and scrutiny as was life on the school grounds. Families were checked beforehand, and certain expectations of living conditions and salaries were elaborated. The school conducted correspondence, questionnaires, and inquiries among friends of the school to investigate the character of the prospective employer and the household. Pratt also wanted to know about the use of alcohol or tobacco, the regularity of church attendance, the kind of people already employed by the host, and whether the host was known to pay his wages on time.[45] During the actual outing, students and hosts needed to send reports to the school, while an "outing agent" toured the worksites, visiting (at least in theory) several times a year.[46]

Before departing to their outing periods, students had to sign a form of conduct, a practical "outing oath," which showcased how white work and performance were perceived in the school. In it they stated:

> Sir:
>
> I want to go out into the country.
>
> If you send me I promise to obey my employer, to keep all the rules of the school.
>
> I will attend Sunday School and Church regularly.

> I will not absent myself from my farm home without permission of my employer and will not loaf about stores or elsewhere evenings or Sundays.
>
> I will not make practice of staying for meals when I visit my friends.
>
> I will not use tobacco nor any spirituous liqueurs in any form.
>
> I will not play cards nor gamble, and will save as much money as possible.
>
> If out for the winter I will attend school regularly and will do my best to advance myself in my studies.
>
> I will bathe regularly, write my home letter every month, and do all that I can to please my employer, improve myself and make the best use of the chance given to me.
>
> Very respectfully,
>
> . . . . Pupil[47]

This "outing oath" provides a window to how Carlisle imagined its Indigenous students should act as white as possible in the white settler world. Obedience, abstinence from vices, and attendance to sanitary measures guided and framed what white work experience was all about, thus determining the parameters of respectable life to which the students should aspire. The "oath" set not only practical guidelines for behavior, but an ideal formula for behaving as white.

While closely regulated, the outing experiences varied in practice. Samuel Kenoi, one of the Apaches who attended Carlisle, remembered that the school was so overpopulated that they sent as many as they could to outings. In 1900, Kenoi worked on a tobacco field and farmed corn, onions, and other vegetables. He also worked as a farmhand, milking cows, working at stalls, and taking care of livestock. Kenoi rejected what the outing offered, as he ran away from his outing assignment after less than a month there. He was dismissed from Carlisle to the reservation soon thereafter, after only one year at the school, the bulk of which he spent on outings.[48] Betzinez, in contrast, cherished his outing experiences. He was sent out during his first school year in April 1888 to the household of a Quaker family, the Coopers, in Bucks County, Pennsylvania. Betzinez described the Coopers as "warm-hearted people" who helped "me to improve myself." Betzinez stayed with the Coopers until 1890, becoming, in his own words, "practically an adopted member of the family." He also worked for other families in the Pennsylvania–New Jersey area, learning the operations of dairy farming as well as raising corn and barley, along with some carpentry, plumbing, and general construction. In all, Betzinez spent a full decade in the books at Carlisle, assigned to outings for approximately three years of that time. Thus, while gaining plenty of work experience on outings, Betzinez actually spent the majority of his time working on campus. At least one of his outings was cut short after a month due to injury, and he worked frequently on campus, doing various types of jobs, including plumbing and general construction, missing class because of it. All this prepared him for a specific role in the settler society as a worker and a farmer, with a varied practical skill reservoir.[49]

That the students had achieved economic independence and equality with other races, being scattered on their own as individuals in the larger society, as Fear-Segal observes, was the ultimate goal of the Carlisle training. Even when he was no longer Carlisle's superintendent, Pratt continued to trumpet the socially transforming powers of the school. Giving a talk in Philadelphia in 1913, he described how "Hundreds of educated Indians have escaped the thralldom of reservation life through non-reservation schools, and have successfully gone into professional and industrial occupations." He also provided examples to prove his point, describing individuals who worked at a saw mill and a locomotive repair shop, and as an automobile painter, a traveling salesman, and a plumber.[50]

In 1910, Carlisle was able to report that more than half of its graduates were making their living away from the reservation, and that out of a total of 514 graduates, only 54 were engaged in farming. Thus, industrial skills had gained the upper hand over farming, which had been the school's early priority for Indigenous futures. Then again, statistics are deceiving, as most students never graduated. Numerous students deserted, or were deemed unfit for work and cultural assimilation by the school and were sent back to their communities.[51]

Conclusion

"The most important aim of the [outing] system was to give the Indian the 'courage of civilization.'" Twenty-one years after Carlisle's closure in 1918, the American educator and professor Robert L. Brunhouse wrote an article for the periodical *Educational Outlook.* In it he evaluated the importance of Carlisle, asserting it had set an example emulated by countless other nonreservation Native American schools in the U.S. He especially raised the outing system as Carlisle's greatest achievement, calling it no less than the "apprenticeship for civilization."[52]

Relying on notions of racial difference, Carlisle promoted industrial training as the quintessential tool for the remaking of Indigenous lives, guiding them toward whiteness and a future in the settler society they would not otherwise have been capable of achieving. In doing so, it acted much like industrial schools did in colonies around the world, making claims that marginalized people would require industrial education to adjust and to survive in a world ruled by whites. At Carlisle, reform meant learning skills—but also working with the whites, joining the settler colonial workforce, and becoming ready to compete with other races in the job market. The outing system in white households and workplaces was applied practice for just that purpose. It was a testing ground. The school shops in turn provided the needed expertise to work with the whites in the first place. Both provided the students the "courage of civilization" Brunhouse mentioned, preparing and uplifting them, and promising entry to the settler society and its respect and status.

In practice, Carlisle amounted to what is today often recognized as cultural genocide among Indigenous activists and communities, as well as scholars. It robbed the Indigenous children from their communities, and systematically subjected them to cultural elimination.[53] While spreading the gospel of work, the training overlooked the wants of the Indigenous students. It also did not fully account for the prejudices of the settler society. Following her discharge from Carlisle in 1895, Ramona Chihuahua worked proudly as the head laundress of the school in Fort Sill, Oklahoma, for three years. She corresponded with her old school as well as her outing family.[54] After leaving Carlisle, Jason Betzinez went to work at the Steelton mill of the Pennsylvania Steel Company, first as a common laborer and then as a blacksmith. Next, he relocated with his Chiricahua kin in Oklahoma. There he found employment as a U.S. Army scout—a job for which Carlisle did not prepare him. He also did varied jobs more suited to his industrial training, mostly as a blacksmith, but also as a carpenter and an artisan.[55]

When asked "did you attended [*sic*] any other schools since leaving Carlisle," Janette Russell replied, "no I didn't work any other school since I came back from Carlisle."[56] Referring to schooling as work, Russell's reply reveals how closely its students linked the Carlisle experience with work. Yet, as did Russell, most Carlisle students returned to their reservation communities, where they encountered poverty, social ills, and dispossession. And, like Russell, they did not carve a future for themselves amid white populations, but worked in varied odd jobs in, or within the vicinity of, the reservation. They put their industrial education to use in a settler society that more often than not did not accept them as equals, and where whites continued to covet their lands and the federal government continued to oversee and curtail their autonomy.

Notes

1. Janette Russell, nee Woods, Student File, 1910, https://carlisleindian.dickinson.edu/student_files/janette-woods-pah-go-stat-un-student-file; Jason Betzinez Student File, 1914, https://carlisleindian.dickinson.edu/student_files/jason-betzinez-bet-zin-ez-student-file; Ramona Chihuahua Student File, 1910, https://carlisleindian.dickinson.edu/student_files/ramona-chihuahua-be-ta-hi-chau-student-file. Unable to defeat the Chiricahua Apaches, the U.S. federal government imprisoned and removed the whole group to Florida as prisoners of war in August–September 1886. From there, the government sent many Chiricahua children and youth to Carlisle in 1886 and 1887, moving the rest of the group again, first to Alabama and then to Oklahoma in 1894, before releasing them in 1913. See Janne Lahti, *Wars for Empire: Apaches, the United States, and the Southwest Borderlands* (Norman: University of Oklahoma Press, 2017); Edwin R. Sweeney, *From Cochise to Geronimo: The Chiricahua Apaches, 1874–1886* (Norman: University of Oklahoma Press, 2010); H. Henrietta Stockel, *Shame and Endurance: The Untold Story of the Chiricahua Apache Prisoners of War* (Tucson: University of Arizona Press, 2004). Janette Russell's first name appears at both Janette and Jeanette in the Carlile files.

2. Matthew Frye Jacobson, *Whiteness of a Different Color: European Immigrants and the Alchemy of Race* (Cambridge: Harvard University Press, 1998), 2–5; David Roediger, *The Wages of Whiteness: Race and the Making of the American Working Class* (New York: Verso, 1991).

3. David Wallace Adams, *Education for Extinction: American Indians and the Boarding School Experience, 1875–1928* (Lawrence: University Press of Kansas, 2020); Jacqueline Fear-Segal, *White Man's Club: Schools, Race, and the Struggle of Indian Acculturation* (Lincoln: University of Nebraska Press, 2007); Brenda J. Child, *Boarding School Seasons: American Indian Families, 1900–1940* (Lincoln: University of Nebraska Press, 1998).

4. *Annual Report Commissioner of Indian Affairs* (hereafter *ARCIA*) *1882, House Ex. Doc. No. 1, 47th Cong., 2d sess. Serial 2100*, 25, 26; *ARCIA 1885, House Ex. Doc. No. 1, 49th Cong., 1st sess. Serial 2379*, 7.

5. There exist numerous histories of specific schools. See, for example, Kevin Whalen, *Native Students at Work: American Indian Labor and Sherman Institute's Outing Program, 1900–1945* (Seattle: University of Washington Press, 2018); K. Tsianina Lomawaima, *They Called It Prairie Light: The Story of Chilocco Indian School* (Lincoln: University of Nebraska Press, 1995); Myriam Vučković, *Voices from Haskell: Indian Students Between Two Worlds, 1884–1928* (Lawrence: University Press of Kansas, 2008); Diana Meyers Bahr, *The Students of Sherman Indian School: Education and Native Identity Since 1892* (Norman: University of Oklahoma Press,

2014); Jacqueline Fear-Segal and Susan D. Rose, eds., *Carlisle Indian Industrial School: Indigenous Histories, Memories & Reclamations* (Lincoln: University of Nebraska Press, 2016). On a transnational frame, see Elisabeth M. Eittreim, *Teaching Empire: Native Americans, Filipinos, and US Imperial Education, 1879–1918* (Lawrence: University Press of Kansas, 2019); Andrew Woolford, *This Benevolent Experiment: Indigenous Boarding Schools, Genocide, and Redress in Canada and the United States* (Lincoln: University of Nebraska Press, 2015); Michael C. Coleman, *American Indians, the Irish, and Government Schooling: A Comparative Study* (Lincoln: University of Nebraska Press, 2007).

6. On settler colonialism being driven by the logic of Native elimination, see Patrick Wolfe, "Settler Colonialism and the Elimination of the Native," *Journal of Genocide Research* 8, no. 4 (2006): 387–409.

7. Genevieve Bell, "Telling Stories Out of School: Remembering the Carlisle Indian Industrial School, 1879–1918" (PhD diss., Stanford University, 1998), 387–389.

8. Richard Henry Pratt, *The Indian Industrial School, Carlisle, Pennsylvania: Its Origin, Purposes, Progress, and the Difficulties Surmounted* (Carlisle: Hamilton Library Association, 1904), 8 and 25.

9. Studies on everyday/intimate colonialism in different colonial situations across the world are too numerous to mention here. See, for example, Ann Laura Stoler, *Carnal Knowledge and Imperial Power: Race and the Intimate in Colonial Rule* (Berkeley: University of California Press, 2002); Tony Ballantyne and Antoinette Burton, eds., *Bodies in Contact: Rethinking Colonial Encounters in World History* (Durham: Duke University Press, 2005); Janne Lahti, *The American West and the World: Transnational and Comparative Perspectives* (London: Routledge, 2019), 116–146.

10. On whiteness and work in the U.S. Empire, see Sarah Steinbock-Pratt, *Educating the Empire: American Teachers and Contested Colonization in the Philippines* (Cambridge: Cambridge University Press, 2019); Janne Lahti, *Cultural Construction of Empire: The U.S. Army in Arizona and New Mexico* (Lincoln: University of Nebraska Press, 2012), 204–212, especially chapters 6 and 7; Katherine Benton-Cohen, *Borderline Americans: Racial Division and Labor War in the Arizona Borderlands* (Cambridge: Harvard University Press, 2009).

11. Daniel E. Bender and Jana K. Lipman, eds., *Making the Empire Work: Labor and United States Imperialism* (New York: New York University Press, 2015); Oliver Charbonneau, "Colonizing Workers: Labor, Race, and U.S. Military Governance in the Southern Philippines," *Modern American History* 4, no. 1 (2021): 25–47; Eittreim, *Teaching Empire*; Rebecca Swartz, *Education and Empire: Children, Race and Humanitarianism in the British Settler Colonies, 1833–1880* (London: Palgrave Macmillan, 2019).

12. J. Kēhaulani Kauanaui and Patrick Wolfe, "Settler Colonialism Then and Now: A Conversation Between J. Kēhaulani Kauanaui and Patrick Wolfe," *Politica & Società* 1, no. 2 (2012): 240.

13. Andrew C. Isenberg and Lawrence H. Kessler, "Settler Colonialism and the Environmental History of the North American West," *Journal of the West* 56, no. 4 (2017): 57–66.

14. On histories of Indigenous labor in U.S. settler society, see David R. M. Beck, *Unfair Labor? American Indians and the 1893 World's Columbian Exposition in Chicago* (Lincoln: University of Nebraska Press, 2019); Brian Hosmer and Colleen O'Neill, eds., *Native Pathways: American Indian Culture and Economic Development in the Twentieth Century* (Boulder: University Press of Colorado, 2004); Colleen O'Neill, *Working the Navajo Way: Labor and Culture in the Twentieth Century* (Lawrence: University Press of Kansas, 2005); Alice Littlefield and Martha C. Knack, eds., *Native Americans and Wage Labor: Ethnohistorical Perspectives* (Norman:

University of Oklahoma Press, 1996); Janne Lahti, "Colonized Labor: Apaches and Pawnees as Army Workers," *Western Historical Quarterly* 39, no. 3 (2008): 283–302.

15. *ARCIA 1880, House Ex. Doc. No. 1, 46th Cong., 3d sess. Serial 1959*, 301; Pratt, *The Indian Industrial School*, 17, 32–33.

16. Fear-Segal, *White Man's Club*, 159; on Pratt's vision, see Adams, *Education for Extinction*, 56–60.

17. Jason Betzinez and Wilbur Sturtevant Nye, *I Fought with Geronimo* (Lincoln: University of Nebraska Press, 1959; reprint, 1987), 152–153.

18. *ARCIA 1880*, 302.

19. Catalogue of the Indian Industrial School, Carlisle, PA., 23d Year, 1902 (no publisher info), 107.

20. Catalogue of the Indian Industrial School, 108.

21. Catalogue of the Indian Industrial School, 108. See also the school's promotional booklet, *Indian Industrial School, Carlisle, PA* (1895), www.loc.gov/item/07030346.

22. "Quotes of Industrial products produced at Carlisle for Indian Service," Carlisle Indian School Digital Resource Center, 1901, https://carlisleindian.dickinson.edu/documents/quotes-industrial-products-produced-carlisle-indian-service.

23. *ARCIA 1882*, 28 (quote); *ARCIA 1880*, 300; Pratt, *The Indian Industrial School*, 32–33 and 43.

24. Catalogue of the Indian Industrial School, 107.

25. Catalogue of the Indian Industrial School, 110.

26. Catalogue of the Indian Industrial School, 108–109.

27. *The Schools News*, July 1881, 3.

28. Industrial Department Output Report for 1900, Carlisle Indian School Digital Resource Center, https://carlisleindian.dickinson.edu/documents/industrial-department-output-1900-1901.

29. Fear-Segal, *White Man's Club*, 182.

30. Janette Woods was sent to outings at F. Wood of Torresdale, Pennsylvania, September 20, 1887–December 16, 1891; J. Cochran of Gettysburg, Pennsylvania, April 14–September 8, 1892; W. Yerkes (Yeokes in Register of Pupils) of Hatboro, April 21–September 13, 1893, and October 12, 1893-Transferred; to F. Wood, Pennsylvania, January 16–October 26, 1895. See Register of Pupils, 1890–1900, Vol 1., 129, 132, 134, 138, 142 (as Jeanette Pahgastatum), Record Group 75, National Archives and Records Administration; Janette Russell Student File; Alicia Delgadillo, ed., *From Fort Marion to Fort Sill: A Documentary History of the Chiricahua Apache Prisoners of War, 1886–1913* (Lincoln: University of Nebraska Press, 2013), 225.

31. Register of Pupils, 1890–1900, Vol 1, 90, 109, 118–119.

32. Ramona Chihuahua Student File.

33. Catalogue of the Indian Industrial School, 11; Adams, *Education for Extinction*, 174.

34. Richard Henry Pratt, "The Advantages of Mingling Indians with Whites," speech given at *Proceedings of the National Conference of Charities and Correction, at the Nineteenth Annual Session held in Denver, Col., June 23–29, 1892*, Carlisle Indian School Digital Resource Center, https://carlisleindian.dickinson.edu/teach/kill-indian-him-and-save-man-r-h-pratt-education-native-americans.

35. Richard Henry Pratt, *Battlefield and Classroom: Four Decades with the American Indian, 1867–1904*, ed. Robert M. Utley (Norman: University of Oklahoma Press, 2003), 283.

36. Jacqueline Fear-Segal and Susan D. Rose, "Introduction," in *Carlisle Indian Industrial School: Indigenous Histories, Memories & Reclamations*, ed. Jacqueline Fear-Segal and Susan D. Rose (Lincoln: University of Nebraska Press, 2016), 6.

37. Catalogue of the Indian Industrial School, 111.

38. Fear-Segal, *White Man's Club*, 173; Robert L. Brunhouse, "Apprenticeship for Civilization: The Outing System at the Carlisle Indian School (1939)," Carlisle Indian School Digital Resource Center, https://carlisleindian.dickinson.edu/publications/apprenticeship-civilizations-outing-system-carlisle-indian-school-robert-brunhouse.

39. Register of Pupils, 1890–1900, Vol 1. 84–143, Vol 2. 84–90.

40. Brunhouse, "Apprenticeship for Civilization," 4.

41. *The Carlisle Arrow*, May 14, 1915, 1. See also *The Carlisle Arrow*, May 7, 1915; March 12, 1915; March 31, 1916.

42. Yearbook, Carlislc, 1917, 13, 21, 24.

43. Yearbook, Carlisle, 1917, 100.

44. On commentaries by former students and hosts, see *The Carlisle Arrow and Red Man, Outing Number*, June 7, 1918, 16–18.

45. Brunhouse, "Apprenticeship for Civilization," 5.

46. Catalogue of the Indian Industrial School, 111.

47. *The Carlisle Arrow and Red Man, Outing Number*, June 7, 1918, 4.

48. Samuel Keno Student File, Carlisle Indian School Digital Resource Center, https://carlisleindian.dickinson.edu/student_files/samuel-keno-student-file.

49. Betzinez and Nye, *I Fought with Geronimo*, 156–159; Jason Betzinez, "My People," 141, Jason Betzinez Collection, Western History Collections, University of Oklahoma Libraries. On Betzinez's outing history at Carlisle, see Jason Betzinez Student File. He worked in Newton, Pennsylvania, Pennington, New Jersey, and Edgewood, Pennsylvania.

50. Richard Henry Pratt, "Negroes and Indians: Address of Brigadier General Richard H. Pratt, U.S.A. Retired, Made before the Pennsylvania Commandery, Military Order of Foreign Wars of the United States," Bellevue Stratford Hotel, Philadelphia, January 14, 1913; Fear-Segal, *White Man's Club*, 175.

51. Fear-Segal, *White Man's Club*, 174–175.

52. Brunhouse, "Apprenticeship for Civilization," 8.

53. A symposium was held in 2012 that brought some 290 people to Carlisle, more than half of whom were Native, representing thirty-six tribal nations. Descendants, activists, musicians, academics, and others shared stories, addressing the histories and legacies of Carlisle in a multigenre setting. See Fear-Segal and Rose, eds., *Carlisle Indian Industrial School.*

54. Delgadillo, *From Fort Marion to Fort Sill*, 48–49; Ramona Chihuahua Student File.

55. Betzinez and Nye, *I Fought with Geronimo*, 201–211; Jason Betzinez Student File.

56. Janette Russell Student File.

CHAPTER 3

Industrial Education and Technical Independence in Spanish American Countries

Helge Wendt

Introduction

In 1856, Domingo Faustino Sarmiento, an exiled Argentine politician and educational modernizer, described the contradictory situation most Spanish American countries faced in the decades following their independence from the Spanish rule: "Modern industry, which is based on primary education, is already knocking on our doors, asking if we know how to build machines and if we have anyone to operate them."[1] According to Sarmiento, on the one hand, national economies needed to open up to foreign investment, but on the other, the former colonial economies of the Americas needed to become self-reliant in an increasingly competitive global market environment.[2] In his writings on the education system, Sarmiento made it clear that schooling at all ages was important for achieving political and economic independence. As in other countries, education would also serve to strengthen the national productive sector.

Industrial schools were a means of achieving these goals. From a comparative perspective, these schools appeared in Chile, Mexico, and Colombia during the nineteenth century, despite different political, regional, and institutional contexts. Their purpose was to train boys and young men, and later women, in practical and technical areas of production, largely in response to the external demands of the international economy, national political changes, and transforming modes of production. This branch of education also had a social and racial component to make poorer and marginalized populations productive parts of the relatively young and diverse national society.

Much of the literature on the history of vocational education in Spanish American countries emphasizes its national character, which underlies its role in building a nation-state, a national economy, and a fairly homogeneous national identity.[3] Some scholars argue that this model of schooling responded to the same principles of reforming education and educational institutions in Europe and the United States, thus overemphasizing the dependence of the peripheries on the global center.[4]

However, Latin American educational institutions were formed within a multidimensional system that emerged from local and global influences. This chapter offers a comparative analysis of industrial schools in nineteenth-century Latin American states. It argues that their emergence and development resulted from transnational connections and knowledge transfers, both in a multilayered context of local and national initiatives and needs, and within interlocal and global transfer of institutional, pedagogical, and epistemic knowledge.[5] In the field of new educational institutions that emerged in the nineteenth century—vocational schools, primary schools, universities, and research institutions, such as observatories and museums—existing educational structures were transformed by a variety of influences, including government initiatives, private associations, and teachers, researchers, and professors of Latin American and European immigrant backgrounds, acting within a context of liberal politics and a positivist philosophical movement. In the Latin American context, industrial schools grappled with the relationship between exploitation and improvement,[6] which is evident in this volume's introduction.

Latin American Vocational Schools in an International Context

In his study of Mexico's technical modernization, Edward Beatty emphasized that industrial schools were intended to contribute to economic development, but their impact was minimal because of weak funding and institutional problems.[7] Comparing the development of industrial vocational schools for boys and young men in different Spanish-speaking American countries during the nineteenth century reveals common features and differences.

There are some fundamental differences between Latin American republics and European models of liberal nation-states, including the immense social, linguistic, and racial diversity of the Spanish-speaking countries of the Americas. This posed a challenge to basic and technical education in Latin American countries and the goal of building homogeneous nation-states.[8] Moreover, the young republics did not promote regional industrialization in the same way as Great Britain, the German countries, or France, for instance.[9] Instead, the most dynamic industries—in the capitalist sense—were related to the exploitation of primary resources, such as sugar production in Cuba, meat production in the Río de la Plata Basin, and the mining sector in the Andean countries and Mexico.[10] Industrial mining and agriculture were part of global capitalist circulation of commodities,[11] local capitalist oligarchic social structures, and increased investments by European and U.S. companies, which in turn intensified their search for a local workforce.[12] However, most of the economic output of the Spanish American republics and the few remaining Spanish colonies of Cuba and Puerto Rico came from small craft enterprises, workshops, and processing plants.[13] Consequently, the meaning of "gospel of work and money" for the pupils of these industrial schools in Spanish America included religious, civilizational, and economic components. Making work a vocation moved between the national and international expectation of progress and the individual limited spaces of social and economic activities.

Within the transformations of political (national) organizations, international relationships, and economic restructuring, conflicts in work culture arose between rural and urban areas, between different social strata, between natives and migrants, and between the capital and the provinces. In the second half of the nineteenth century, these countries experienced a radical transformation in some sectors of production, with the establishment of large factories that began producing commodities on a large scale, and integrating into global material flows and economic transactions.[14] The newly founded vocational schools were expected to lead to the creation of a technology-based industrial production or to help the graduates start small businesses.[15] Pupils learned to build small machines, fabricate tools, and repair imported machinery.[16] As a result, in addition to fantasies of European-style nation-building and industrialization, this newly formed skilled labor force was expected to reduce dependence on importing new machinery. In these activities, vocational schools played a vital role at the local level, beyond national and international aspirations.[17] They also provided skills and knowledge that were used to fabricate handicraft products, often of local significance and distribution. In addition, some of these vocational schools evolved into schools of engineering and natural sciences, or became universities, such as the University of Antioquia in Medellín, Colombia.[18] However, the interrelated transformations in the areas of higher education, institutionalization, and engineering are beyond the scope of this comparative overview, which focuses on vocational training institutions for pupils between the ages of six and about twenty-five.

Based on the experiences from their home countries or during their education abroad, immigrated educators contributed to the development of these schools, research institutions, laboratories, and curricula. Well-educated *criollos* (descendants of Spaniards born in the Americas) promoted the global character of these institutions by establishing connections to European countries. They mobilized knowledge, objects, students, and teachers in educational exchanges. Schools adapted to changing circumstances and paid particular attention to the work opportunities of their students when designing their curricula.

In addition to the flow of materials, educational and pedagogical programs, as well as teachers between Spanish America and Europe, there was also considerable exchange between the republics, for example, between Argentina and Chile or Venezuela and Chile.[19] The politicians and educators Domingo Faustino Sarmiento and Andrés Bello passionately promoted vocational education as a means for unstable and economically weak republics to participate in a prosperous modernity.[20]

These institutions and developments should be analyzed in a comparative and interconnected way.[21] The comparative perspective was already invoked by actors of the time, such as Domingo Faustino Sarmiento. In a report to the Chilean Council of Universities in 1856, he compared the situation in Chile with other countries in South America, the United States, France, Prussia, and England, as models for reforming the Chilean educational sector. However, he expected that the educational institutions he proposed would respond to the specific national contexts.

The general view on the history of vocational training in Spanish America during the second half of the nineteenth century offers some points of comparison. First, the context of the creation of such schools, and, second, the history of their development, as well as the transformations of the curricula and the recruitment of pupils of different ages, offer comparable points of view. A third point of comparison is the international connectedness and the national framework of these institutions. A final comparison examines the different national trajectories in relation to how these schools were integrated within the respective national economies. Following an intersectional approach to the issue of industrial schools in some Latin American states during the nineteenth century, a multilayered placement in local, regional, national, and international contexts is the basis of the following overview.

Chile

Reform efforts in Chile date back to the 1800s. The Creole lawyer and politician Manuel Salas, before independence, ensured that mathematics and technical drawing were taught to strengthen the productive sector.[22] The creation of the Instituto Nacional, which taught mathematics and drawing as well as sciences and classics, is associated with two founding fathers of the Republic of Chile, the politician and philosopher Juan Egaña and the priest and politician Camilo Henríquez. In this institution, however, little practical or technical knowledge was taught. In addition to many Chilean teachers, some immigrant instructors were hired, such as the French military geographer Carlos Ambrosio Lozier, who taught mathematics and French and became the fourth director of the Instituto Nacional in 1825. He had emigrated to the United States after Napoleon's defeat, tried his luck in Brazil and Buenos Aires, and then became a member of various government topographic commissions in Chile.[23] Finally, Lozier was also involved in the founding of the first industrial school in Chile in 1822, whose goal was to enable Chilean entrepreneurs to extract and utilize ores more effectively. In this school, Camilo Henríquez encouraged the teaching of subjects such as mechanics, chemistry, agriculture, and industrial production.[24] The National Institute in Santiago de Chile, on the other hand, was designed to prepare pupils for university studies, as was a similar institution in Mexico.[25]

Higher education in Chile was based on the international exchange of different types of knowledge. For instance, the Pedagogical Institute of Santiago was founded after Chilean professors consulted with various German experts and visited German pedagogical colleges.[26] Professionalized teacher training also increased the number of students, so that by 1900, there were about 170,000 students enrolled in public and private primary schools. The total number of pupils in secondary education was about 12,000. In her work on Chilean educational history, Amanda Labarca counted only 1,230 pupils studying in more than twenty different technical schools and institutes,[27] showing that higher and specialized education in Chile was not a mass phenomenon at the turn of the twentieth century. A new reform of the educational system began in 1912 with the aim of introducing more technical subjects into the humanities-dominated education.[28]

One of the most important schools of technical education was the School of Arts and Trades (Escuela de Artes y Oficios) in Santiago. The school was founded in 1843 by the Agricultural Society of Santiago and entrusted to Baron Auguste Picolet d'Hermillon, the son of the Sardinian ambassador to Buenos Aires and Montevideo. Picolet d'Hermillon apparently based the curriculum on that of the Evening School of Design in Turin.[29] In addition to the School of Arts and Trades, the Agricultural Society also founded a School of Agronomy and later a School of Veterinary Medicine.[30]

The creation of vocational schools began a tradition in Chile that was revived by other associations promoting industrialization policies and industrial education. One of them was the Sociedad de Fomento Fabril (Sofofa), which, during the 1880s, founded a professional school for women, a school for electrical workers, a technical-commercial institute, and the Drawing School of Valparaíso. Finally, the newly established Institute of Engineers was meant to reorganize and structure the industrial development of Chile in the aftermath of the Civil War of 1891.[31]

The School of Arts and Trades opened its doors in 1849, with French-born Julio (Jule) Jariez as its first director.[32] The four-year program included courses in technical and industrial drawing, and practical training in the workshops. In 1851, evening classes in arithmetic, geometry, and applied drawing were introduced, and another School of Arts and Trades was established in Talca.[33]

In 1864, the government redefined the requirements of this school. The guidelines stated in the first article: "The purpose of the School is to train educated, industrious and honest artisans who will contribute to the progress of industry in Chile."[34] In an effort to include other regions in a supposedly monocultural and hierarchical system,[35] the regulations stipulated that the students must be between sixteen and twenty-one years old and recruited from the different provinces of Chile. In addition, only an elementary level of Spanish was required, which was appropriate considering the country's large immigrant and Amerindian population—although only Catholics were accepted. The examinations were held in the regional capitals and tested prospective students on their knowledge of arithmetic and the metric system. Examining commissions were set up for this purpose, which included an appointed supervisor (*visitador*). Upon admission, the students made a commitment to work in the profession for which they had been trained. Students at the School of Arts and Trades were free to choose their specialty in the forge, the smeltery, the boiler shop, or one of the various woodworking shops. In addition, the director assigned students to the mechanical workshops according to their previous knowledge.[36]

At the School of Arts and Trades at Santiago, theoretical classes included arithmetic, algebra, geometry, trigonometry, Spanish grammar, catechism, geography, history of Chile, writing, and basic physics and chemistry. Drawing classes, which were closely related to technical understanding and practical application in the workshops, included the following teaching content: "linear and ornamental geometry, descriptive geometry, machines, buildings, construction projects, water brush sketches, and, in general, the practice of those graphic procedures and problems of application that correspond more directly to each student's workshop."[37]

Like these more theoretical subjects, several practical subjects were taught over the course of four years. These included instruction in seven workshops: mechanics, iron forging, smelting, copper kettle forging, plumbing, model making, carpentry, carriage building, and other forms of woodworking. In 1893, the school moved to a newly constructed campus in Santiago de Chile.[38] In order to cope with the uneven regional industrial development and to meet the expectations and needs of different industries, other industrial schools were established in Concepción and Chillán.[39]

Vocational schools for women were established in the 1870s, with the opening of the first female vocational schools in Valparaíso and Copiapó.[40] These schools were reformed in the 1880s, renamed "Schools of Arts and Trades for Women," and foreign-born teachers were hired to expand the curriculum.[41] The decree of January 9, 1887, listed the subjects to be taught: bookkeeping, various areas of dressmaking, leatherwork, kitchen management, laundry and ironing, and the basics of geometric drawing. Pupils needed to be above the age of twelve and have an elementary-school education. As in the boys' schools, drawing was considered the foundation of the girls' education for later professional practice.[42]

Vocational schools in Chile, especially the Schools of Arts and Trades, were designed to train young men and women to handle machines and to run workshops. The first objective of vocational schools was to enable pupils to copy or adapt imported machines and tools to the local market. The second goal was to integrate graduates into the local economic structures, either in workshops or in public enterprises, in railway and maritime companies, or, at the turn of the twentieth century, in electricity companies.[43]

Mexico

As in Chile, the *Escuelas de artes y oficios* (Schools of Arts and Trades) in Mexico were intended to provide a systematic approach to vocational training.[44] They were designed to strengthen Mexico's industrialization efforts through substantial and rigorous training programs. Another goal was to combat poverty in the different Mexican states.[45] Authors such as Sonia Pérez Toledo and Carlos Illade have related the role of the working class to the process of state formation in Mexico during the nineteenth century.[46] Among the private initiatives to establish educational institutions in the period around 1800 were the projects of Miguel Hidalgo, the former priest and leader of Mexican independence. Also, the Escuela Patriótica (Patriotic School), which taught practical knowledge, crafts, and agriculture, was dedicated to the education of orphans and their "integration" into society.[47] Another initiative within this prehistory of vocational schools in Mexico was the manifesto of the Mexican politician and author Lorenzo de Zavala, titled: "The Utility of a National Society of Agriculture and Industry, Regarding a Rural School and Another of Arts and Trades of the Mexican Republic." Under the pseudonym "Friend of the People," the ardent federalist Zavala called for the creation of an independent Mexican production-based economy that would include all social and ethnic groups. This utilitarian vision included the Indigenous population and suggested that the training of skilled laborers would help elevate Mexican industry.[48]

In addition to the projects and schools established during the Napoleonic wars and early independence, the paternalistic initiatives of the Lancaster Association in Mexico aimed to integrate the poor social classes into productive parts of society.[49] Based on the peer tutoring system that Joseph Lancaster had established in his schools in Great Britain, the Association opened a school in Mexico City in 1822.[50] However, the school's pedagogical plan of older pupils mentoring younger students was not implemented until the 1830s. Other schools following the Lancasterian method of education were opened in Guadalajara and Jalisco, or were unsuccessfully planned, as in Chihuahua.[51] These schools concentrated on teaching the basics of mathematics, reading, and writing, while technical education was of secondary importance.

The French-born Eduardo de Turreau de Linères was one of the most active proponents of idealistic plans to create a strictly technical school.[52]

In Mexico, vocational schools were established in some of the state capitals. In 1842, the governor of Jalisco ordered the establishment of a school of mechanical arts that was later transformed into a School of Arts and Trades.[53] Another school was founded in Puebla (de los Ángeles), which was already an important center of textile and iron production.[54] A government decree of 1843, which was not implemented, provided for the establishment of a school in Mexico City. Despite its failure, the planned program established some principles for later school foundations. Behind the initiative was Lucas Alamán, an entrepreneur and political leader, who promoted the mechanization of Mexican production.[55] The decree sought to establish a boarding school in Mexico City to train pupils in various trades. The curriculum was divided into four main components: drawing, applied mathematics, applied chemistry, and applied mechanics. Practical lessons focused on smelting and alloying, metal works and woodworking, and linen processing. It was envisioned that practical lessons in gilding, locksmithing, and saddlery would be added at a later date. As in Chilean schools, a basic knowledge of reading, writing, and arithmetic was required. Finally, there was a plan to send four particularly gifted students to Europe each year for further education.[56]

The second initiative for establishing a technical school was successfully implemented by the government of Ignacio Comonfort in April 1856. With the help of special funds, pupils from different Mexican states and territories were recruited to live on the campus, where several workshops were to be built.[57] José Urbano Fonseca, an experienced clerk and academician, became the first director in 1856 and immediately presented a law stipulating that students should be between thirteen and sixteen years old, be physically healthy and of moral integrity, and be able to read and write, but not necessarily know mathematics.[58] A decree of 1857 regulated the offices and detailed the subjects to be taught. A mechanical engineer was hired as the director of technical education, and individual teachers were in charge of specific workshops in ironwork, carpentry, carriage building, stonemasonry, saddlery, shoemaking, and tailoring.

Classes were divided into beginner and advanced levels, as well as general and specialized lessons. The beginners were taught arithmetic, geometry of surfaces, and the respective applications of these two subjects in different trades. In the second year, algebra, geometry, trigonometry, physics, industrial mechanics, and chemistry for practical applications were taught.[59] Finally, a decree by President Comonfort secured the financial situation of this institution and the planned construction activities.[60] In its fight against poverty, the school taught a total of 150 students, a third of whom were considered poor or disadvantaged.[61] In the 1880s, the Mexican government decided to transform the purely professional schools into penitential institutions, where education and military-style discipline would convince delinquent young men to leave their criminal past behind.[62]

Vocational training for young women had some precursor institutions in Coahuila and Chiapas. The state encouraged the training of female teachers so that by the early twentieth century, about two-thirds of Mexican elementary-school teachers were women. In 1871, during the last presidency of Benito Juárez, the Female School of Arts and Trades was founded. As in Chile, this professionalization coincided with the fact that some of the branches of production demanded a high amount of manual labor. Vocational training in sectors traditionally dominated by women facilitated their integration into the labor force.[63] The impact of vocational schools on improving women's skills and working conditions was rather limited, due to the assumption of gender roles and an economic system in which the employment of large masses of rather poorly educated workers was the norm.[64]

A major reform of vocational training took place under Porfirio Díaz, who changed the admission requirements of the School of Arts and Trades in 1877. Pupils had to be no older than twelve and have a solid elementary-school education. After three years of lessons at the school, students would become full apprentices with a regular salary.[65] Although many vocational schools were established under the control of the federal or state government, many students left school early and the number of graduates was quite low. The school in Puebla was founded in the late 1880s, with a rigorous four-year curriculum. It had over a hundred pupils who were introduced to a wide range of arts. In addition to work with classical materials (wood, leather, textiles), they were trained in more recent technical developments such as gas lighting, galvanization, and telegraphy. The National School of Engineers (Escuela Nacional de Ingenieros)—the former Mining School of Mexico—specialized in subjects related to electricity, but in 1889 changed its focus to teach engineers and develop telecommunication and the electrification of production rather than simply train a skilled workforce.[66] In the decades that followed, the pedagogy and the subjects were reformed several times, but the general goal of improving the technical skills of workers and subsequently the national industry remained the same.[67]

In a new reform effort, President Porfirio Díaz sent Félix Palavicini, the headmaster of the vocational-technical annex to the National School of Education, to Europe and the United States to observe the pedagogics employed there. Palavicini proposed the introduction of technical education for boys and girls at an early age in order to improve the workforce and, as part of a clientelist political maneuver, to satisfy the labor unions on the eve of the Mexican Revolution.[68] After the revolution, funding for the public-school sector increased, which supported and expanded elementary and vocational schools.[69]

The development of vocational schools in Mexico in the second half of the nineteenth century was marked by the state's efforts to train workers who could develop new economic sectors in Mexico. In the process, the state ensured that basic principles—such as Catholic faith, loyalty to the state, and moral behavior—were respected. Throughout the various and sometimes violent changes of government, the Schools of Arts and Trades were established in the country's major cities, but they remained only isolated centers of systematic vocational training in the trades.

Colombia

During the decades of devastating civil wars in Colombia between 1830 and 1870,[70] several vocational schools emerged throughout the country.[71] The founding of the school in Medellín, as well as several satellites in smaller cities, is part of a transnational history of educational system reform that relied on the recruitment of French and German teachers. At the same time, these Colombian institutions, founded in the 1860s, were characterized by their grounding in regional production traditions.[72] Some of them, as in Bogotá and Medellín, had the financial resources to offer higher education to their students. Institutes in smaller towns, however, were limited to teaching trades that were already important for the regional economy.

The school in Medellín, founded in 1869, was closely tied to the local mining industry. It was so well equipped that it could manufacture and sell its own tools and small machines,[73] and quickly became an institute affiliated with the University of Antioquia that had been established shortly earlier. The Colombian Military College (founded in 1848), whose mission was to train military and civil engineers, and the Engineering School of Bogotá (founded in 1867) had a similar educational goal. At these institutions, several mathematicians who had studied in France taught mathematics using translated French textbooks.[74]

Admission requirements for the School of Arts and Trades in Medellín stipulated that pupils should be between twelve and eighteen years old, and be baptized, or at least have good knowledge of the Catholic catechism. Religious instruction was part of the curriculum, as in Chile and Mexico, but lack of knowledge in this area could be compensated for if technical talent was evident.

As in the other schools, the "morals" and physical condition of prospective pupils were considered. Finally, the candidates were expected to be literate and to have studied (among other things) Spanish grammar, the first four rules of arithmetic, and some metrology. During their four years of study, the pupils learned mathematics, Spanish, drawing, and some basic knowledge of physics and chemistry. Practical classes included mechanics, ironwork, smelting, modeling, carpentry, cabinetmaking, locksmithing, boiler making, plumbing, and carriage making.[75] Some of the staff at the school was foreign, including the German director, Enrique Haeusler. Originally an engineer, Haeusler came to Antioquia in 1839 and constructed bridges, built a reaction ferry, and worked on the first minting machine at the Casa de Moneda de Medellín, along with Eugenio Lutz, a French-Swiss engineer, who became professor of drawing and practical geometry at the school.[76]

As in Mexican and Chilean schools, the curriculum was modernized to include telegraphy, shorthand, industrial physics, and chemistry. The school had a considerable moralizing mission and was supposed to help educate the working class to become better citizens.[77] The first article of the school's regulation stated: "the School of Arts and Crafts of the Sovereign State of Antioquia, has at its aims the training of educated, industrious and honest artisans, who, by their example and with their knowledge, will contribute to the advancement of industry in the State, and to the reform of our working classes."[78]

Comparing Vocational Schools in Latin America

On the eve of the First World War, the educational systems of most Spanish American countries were reformed and the number of pupils enrolled in vocational schools increased significantly. The last two decades of the nineteenth century witnessed an expansion of school activities in these countries, particularly the growth of elementary schools, but also an increase in vocational schools. The establishment of schools dedicated to the vocational training of girls also played a role here, but some were short-lived. Schools in Mexico City, Medellín, and Santiago de Chile were transformed into university-like institutions. Another Mexican school, in Cuajimalpa, became part of the Universidad Autónoma Metropolitana in order to complete the study program.[79]

The foundation of the first Schools of Arts and Trades was related to the last phase of colonial reforms, when the creole elites started to adopt a more nationalist perspective.[80] In Mexico and Chile, centralization took place in the respective capitals. Pupils were recruited from different provinces, states, or territories with the goal of industrially developing every part of the country and creating a nationwide homogeneous citizenship. In order to support these missions, examinations of primary school students were organized with local committees, a structure that was replicated across the Spanish American countries. In addition, by comparing the creation of the same type of school in different cities, their relatedness to local interests and dynamics in the educational, political, and industrial sectors can be determined.

Another comparable feature of the Schools of Arts and Trades was the admission of teenaged pupils, but since foundational knowledge was a prerequisite, the pool of candidates was rather limited. Another requirement for admission was the "morality" of the candidates, even though the primarily practical education offered no guidance for good behavior except for emphasizing the value of industriousness. This problem was often solved in subsequent reform initiatives by a quasi-military form of organization.

The schools were tied to national and local horizons of expectation. However, foreign teachers, often French immigrants, were hired as founding teachers, bringing their experience from the *Écoles d'arts et métiers*. The French educators' knowledge of French textbooks, some of which they probably helped to translate into Spanish, was an important factor in a transatlantic transfer of knowledge within the field of technical education.

Another element that contradicts the purely national character of this type of school is the heterogeneity of the students. Since their linguistic, cultural, and social backgrounds were diverse, those in charge found it necessary to resort to religious education or military discipline as a medium of homogenization. Further, graduates often chose professions or working environments that did not necessarily coincide with the educational and economic goals of the schools, and so contributed much less to national development than expected. In the early twentieth century, many schools were transformed into university institutes, with a different approach to technical education focused on engineering.[81]

The differences in the industrialization of Latin America compared to European countries are also reflected in the history of technical schools. They were subject to a constant search for funding, official support, and educational programs. In addition, the schools had to prove themselves in an economic context that was highly volatile and offered little consistent orientation. Comparing various Latin American countries allows us to identify similar developments that break through respective national specificities. Further research is needed to explore the extent to which this developmental step, which occurred at the beginning of the twentieth century, was also a result of international connections and transfer of knowledge.

Notes

1. Domingo Faustino Sarmiento, *Memoria sobre educacion comun presentatda al Consejo Universitario de Chile* [Report about primary education presented to the University Council of Chile] (Santiago de Chile: Imprenta del Ferrocarril, 1856), 192–193. All English translations of non-English-language titles of primary source material and quotes were conducted by the author.

2. Edward Beatty, *Technology and the Search for Progress in Modern Mexico* (Oakland: University of California Press, 2015), 193–196; John Tutino, *From Insurrection to Revolution in Mexico: Social Bases of Agrarian Violence, 1750–1940* (Princeton: Princeton University Press, 1984).

3. Mark D. Szuchman, "Childhood Education and Politics in Nineteenth-Century Argentina: The Case of Buenos Aires," *Hispanic American Historical Review* 70, no. 1 (1990): 109–138; Rodrigo Hernán Torrejano-Vargas y Henry Bocanegra-Acosta, "En búsqueda de la identidad social del ingeniero: una trayectoria de construcción de identidad profesional en Colombia en medio de la modernizacióncapitalista (1848–1929)" [In search of the social identity of the engineer: a trajectory of professional identity construction in Colombia in the midst of capitalist modernization (1848–1929)], *Revista eleuthera* 19 (2018): 15–34.

4. Frank Safford, *The Ideal of the Practical: Colombia's Struggle to Form a Technical Elite* (Austin: University of Texas Press, 1976), 55–56.

5. I take the idea of the multilayered transfer of packages of knowledge from Jürgen Renn, "Survey: The Place of Local Knowledge in the Global Community," in *The Globalization of Knowledge in History*, ed. Jürgen Renn (Berlin: Max-Planck-Gesellschaft zur Förderung der Wissenschaften, Edition Open Access, 2012), 10.34663/9783945561232-20.

6. Mary Kay Vaughan, "Primary Schooling in the City of Puebla, 1821–60," *Hispanic American Historical Review* 67, no. 1 (1987): 39–62.

7. Edward Beatty, *Technology and the Search for Progress in Modern Mexico* (Oakland: University of California Press, 2015), 195–196.

8. Charles A. Hale, *The Transformation of Liberalism in Late Nineteenth-Century Mexico* (Princeton: Princeton University Press, 1989), 3, 233, 245; John Jairo Cardenas-Herrera, "The Invention of the Classroom: Republic, Citizenship, and School Space in Colombia: 1819–1830," *Paedagogica Historica* 57, no. 1–2 (2021): 39–49; Juan Mansilla Sepúlveda et al., "Instalación de la escuela monocultural en la Araucanía, 1883–1910: dispositivos de poder y Sociedad Mapuche," [Installation of the monocultural school in Araucanía, 1883–1910: power devices and Mapuche society], *Educação e Pesquisa* 42, no. 1 (2016): 213–228.

9. C. A. Bayly, *The Birth of the Modern World, 1780–1914: Global Connections and Comparisons* (Oxford: Wiley-Blackwell, 2003), 167–186; Sebastian Conrad, *Globalgeschichte: Eine Einführung* [Global history: An introduction] (Munich: C. H. Beck, 2013), 105–106.

10. Alan Dye, *Cuban Sugar in the Age of Mass Production: Technology and the Economics of the Sugar Central 1899–1929* (Stanford: Stanford University Press, 1998); Saul Guerrero, *Silver by Fire, Silver by Mercury: A Chemical History of Silver Refining in New Spain and Mexico, 16th to 19th Centuries* (Leiden: Brill, 2017); Sandra Kuntz Ficker, "De las reformas liberales a la Gran Depresión, 1856–1929," [From liberal reforms to the Great Depression, 1856–1929], *Historia económica general de México: de la colonia a nuestros días* [General economic history of Mexico: from the colony to the present day] (Mexico: El Colegio de Mexico, Secretaría de Economía, 2010), 305–352; Lucía Lewowicz, *LEMCO: Un coloso de la industria cárnica en Fray Bentos, Uruguay* [LEMCO: A colossus of the meat industry in Fray Bentos, Uruguay] (Montevideo: INAC, 2016).

11. Jürgen Osterhammel, *The Transformation of the World: A Global History of the Nineteenth Century*, trans. Patrick Camiller (Princeton: Princeton University Press, 2014); Gregory T. Cushman, *Guano and the Opening of the Pacific World: A Global Ecological History* (Cambridge: Cambridge University Press, 2013).

12. Stephen H. Haber, *Industry and Underdevelopment: The Industrialization of Mexico, 1890–1940* (Stanford: Stanford University Press, 1989); Mario Cerutti, *Propietarios, empresarios y empresa en el norte de México. Monterrey: de 1848 a la globalización* [Owners, entrepreneurs and companies in northern Mexico. Monterrey: from 1848 to globalization] (Mexico: Siglo Veintinuno Editores, 2000); Manuel Llorca-Jaña and Juan Navarrete-Montalvo, "The Chilean Economy During the 1810s–1830s and Its Entry into the World Economy," *Bulletin of Latin American Research* 36, no. 3 (2017): 354–369.

13. Peter N. Stearns, *The Industrial Revolution in World History* (Boulder: Westview Press, 2013), 165.

14. Enrique Cárdenas Sánchez, *Cuando se originó el atraso económico en México. La economía mexicana en el largo siglo XIX, 1780–1920* [When the economic regression originated in Mexico. The Mexican economy in the long 19th century, 1780–1920] (Madrid: Editorial Biblioteca Nueva, Fundación José Ortega y Gasset, 2003), 141–235; Kuntz Ficker, "De las reformas liberales a la Gran Depresión"; Carmen Cariola Sutter y Osvaldo Sunkel, *La historia económica de Chile, 1830 y 1930: dos ensayos y una bibliografía* [The economic history of Chile, 1830 and 1930: two essays and a bibliography] (Madrid: Ed. Cultura Hispánica del Instituto de Cooperación Iberoamericana, 1983); Bárbara de Vos Eyzaguirre, *El surgimiento del paradigma industrializador en Chile (1875–1900)* [The emergence of the industrializing paradigm in Chile (1875–1900)] (Santiago de Chile: Centro de Investigaciones Diego Barros Arana de la Dirección de Bibliotecas, Archivos y Museos, 1999).

15. Amanda Labarca, *Historia de la Enseñanza en Chile* [History of education in Chile] (Santiago de Chile: Imprenta Universitaria, 1939); Estela Eguiarte Sakar, *Hacer ciudadanos. Educación*

para el trabajo manufacturero en el s. XIX en México [Make citizens. Education for manufacturing work in the 19th century in Mexico] (Mexico: Universidad Iberoamericana, 1989); Juan José Saldaña, "Sobre la historia política de la ciencia y la tecnología en México" [On the political history of science and technology in Mexico], *Conocimiento y acción. Relaciones históricas de la ciencia, la tecnología y la sociedad en México* [Knowledge and action. Historical relationships of science, technology and society in Mexico] (Mexico: Facultad de Filosofía y Letras, UNAM, 2011), 21; Juliana Álvarez Olivares, "La Escuela de Artes y Oficios de Medellín y la profesionalización de los artesanos. 1869–1901" [The School of Arts and Crafts of Medellín and the professionalization of artisans. 1869–1901], *History y Sociedad* 26 (2014): 99–119; Eduardo Castillo, "Artesanos, técnicos e ingenieros: La Escuela de Artes y Oficios de Santiago, EAO" [Craftsmen, technicians and engineers: The School of Arts and Crafts of Santiago, EAO], *Atenea* 511, no. 1 (2015): 247–256.

16. Zenobio Saldivia M. and Griselda de la Jara N., "La Sociedad Nacional de Agricultura en el siglo XIX chileno: su rol social y su aporte al desarrollo científico-tecnológico" [The National Agricultural Society in 19th century Chile: its social role and its contribution to scientific-technological development], *Scripta Nova. Revista Electrónica de Geografía y Ciencias Sociales Universitat de Barcelona* 5 [Scripta Nova. Electronic journal of geography and social sciences University of Barcelona 5] (2001): 79–104.

17. In an attempt to overcome the classical narrative of center and periphery in the production of knowledge, see the article on the role of Ecuador in global networks of knowledge production by Elisa and Anna Sevilla, "Inserción y participación en las redes globales de producción de conocimiento: el caso del Ecuador del siglo xix" [Insertion and participation in global knowledge production networks: the case of 19th century Ecuador], *Historia Crítica* 50 (2013): 79–103.

18. Libertad Díaz Molina and Juan José Saldaña, "El papel del Estado y de los ingenieros en el proceso de instauración de políticas públicas en la industria eléctrica en México" [The role of the State and engineers in the process of establishing public policies in the electrical industry in Mexico], *Relaciones históricas de la ciencia, la tecnología y la sociedad en México* (Mexico: Facultad de Filosofía y Letras, UNAM, 2011), 151–182; Claudio Gutiérrez y Michael Reynolds, "Institucionalización y profesionalización de la Ingeniería en Chile" [Institutionalization and professionalization of Engineering in Chile], *Quipu* 16, no. 1 (2014): 63–86; Andrea Flores Hernández, "Ingeniería e ingenieros en la historiografía chilena," *Historia* 396, no. 2 (2017): 463–480.

19. Manuel Lucena Giraldo, "Filantropos u oportunistas? Ciencia y política en los proyectos de obras públicas del Consulado de Cartagena de Indias, 1795–1810" [Philanthropists or opportunists? Science and politics in the public works projects of the Consulate of Cartagena de Indias, 1795–1810], *Revista de Indias* 52, no. 195–196 (1992): 627–646.

20. Sarmiento, *Memoria sobre educacion comun presentatda al Consejo Universitario de Chile.*

21. Michel Espagne y Michael Werner, "Deutsch-französischer Kulturtransfer im 18. und 19. Jahrhundert. Zu einem neuen interdisziplinären Forschungsprogramm des CNRS" [German-French cultural transfer in the 18th and 19th centuries. On a new interdisciplinary research program of the CNRS], *Francia* 13 (1985): 502–510.

22. Domingo Amunátegui y Solar, *Los primeros años del Instituto nacional (1813–1835)* [The first years of the Instituto Nacional (1813–1835] (Santiago de Chile: Imprenta Cervantes, 1889); Claudio Gutiérrez, *Educación, ciencias y artes en Chile (1797–1843): revolución*

y contrarrevolución en las ideas y políticas [Education, sciences and arts in Chile (1797–1843): revolution and counterrevolution in ideas and policies] (Santiago de Chile: RiL editores, 2011).

23. Margaret V. Campbell, "Education in Chile, 1810–1842," *Journal of Inter-American Studies* 1, no. 3 (1959): 357.

24. Camilo Henríquez, "El Instituto nacional" [The national institute], *El Mercurio (de Chile)* 1, no. 15 (1822): 311–312.

25. Fernando Campos Harriet, *Desarrollo educacional, 1810–1960* [Educational development, 1810–1960] (Santiago de Chile: Editorial Andrés Bello, 1960), 53.

26. Carlos Sanhueza Cerda, "La gestación del Instituto Pedagógico de Santiago y la movilidad del saber germano a Chile a finales del siglo XIX" [The creation of the Pedagogical Institute of Santiago and the mobility of German knowledge to Chile at the end of the 19th century], *Estudios Ibero-Americanos* 39, no. 1 (2013): 54–81.

27. Labarca, *Historia de la Enseñanza en Chile*, 216–217. Labarca included around 630 pupils at the Escuela de Sordo-Mudos, Escuela de Ciegos, Conservatorio Nacional de Música, and the Escuela de Bellas Artes. However, I deduct these students from the total number she gives of 1,860 pupils as they did not completely fulfill the criteria of technical and industrial education. The statistic was reproduced by Carmen Gloria Núñez Muñoz, "Educación media técnico profesional en Chile. Una educación para los otros" [Professional technical secondary education in Chile. An education for others], *Acta Académica* (2009): 3.

28. Rolf Arnold and Ulrich Krammenschneider, "Caught Between Private and Public Interests: Vocational Training in Chile," *International Journal of Sociology* 29, no. 2 (1999): 10–11.

29. Labarca, *Historia de la Enseñanza en Chile*, 125.

30. Saldivia and de la Jara, "La Sociedad Nacional."

31. Vos Eyzaguirre, *El surgimiento del paradigma*, 60–65; Jaime Parada Hoyl, "La profesión de ingeniero y los Anales del Instituto de Ingenieros de Chile, 1840–1927" [The engineering profession and the Annals of the Institute of Engineers of Chile, 1840–1927], *Anales del Instituto de Ingenieros de Chile: Ingeniería y Sociedad 1889–1929* [Annals of the Institute of Engineers of Chile: Engineering and Society 1889–1929] (2011): 9–77.

32. Jariez had been a teacher at different Schools of Arts and Offices in France and was vice-director of such a school in Aix-en-Provence. "Jules Jariez," Bibliothèque patrimoniale numérique. https://patrimoine.ensam.eu/actualites/article/8-jules-jariez. He was the author of several pedagogical works on the teaching of mechanics in French, which were later translated into Spanish. Claudio Gutiérrez Gallardo y Flavio Gutiérrez Albornoz, "Física. Su trayactoria en Chile (1800–1960)," *Historia (Santiago)* 39, no. 2 (2006): 485.

33. Labarca, *Historia de la Enseñanza en Chile*, 159.

34. Original text: "La Escuela tiene por objeto formar artesanos instruidos, laboriosos i honrados que contribuyan al progreso de la industria en Chile." Supremo Gobierno, "Reglamento para la Escuela de Artes i Oficios dictado por el Supremo Gobierno el 22 de enero de 1864" [Policy of the School or Arts and Trades, given by the main government, January 22, 1864] (1864), 3.

35. Sepúlveda et al., "Instalación de la escuela monocultural en la Araucanía," 217.

36. Supremo Gobierno, "Reglamento para la Escuela de Artes," 22–23.

37. Supremo Gobierno, "Reglamento para la Escuela de Artes," 25.

38. Photographs from the early twentieth century can be found at: http://lugaresdeciencia.cl/ejes/industria/escuela-de-artes-y-oficios.

39. "La estadística del departamento de Curicó" [The statistics of the department of Curicó], *Boletín de la estadística industrial de la República de Chile 1894–1895* no. 2 [Bulletin of industrial statistics of the Republic of Chile 1894–1895 no. 2] (1895). Agustín Muñoz Sierpe, *Enseñanza universitaria y técnica de Chile. Exposición de Quito (1909: Ecuador). Pabellón de Chile* [University and technical education in Chile. Exposition and Quito (1909: Ecuador). Pavillion of Chile], 1909, 24.

40. Labarca, *Historia de la Enseñanza en Chile*, 164–166; Paulina Dittborn Cordua, "Historia y perspectivas acerca de la educación técnica de nivel superior" [History and perspectives on higher level technical education], *Calidad en la Educación* 27 (2007): 18–33.

41. Labarca, *Historia de la Enseñanza en Chile*, 203–210.

42. José Manuel Balmaceda, "Crónica: Escuela de Artes i Oficios para Mujeres" [Chronicle: School of Arts and Trades for Women], *Boletín de la Sociedad de Fomento Fabril* 5, no. 1 (1888): 43–44.

43. Castillo, "Artesanos, técnicos e ingenieros," 249–252.

44. María de Lourdes Herrera Feria, "La sistematización del saber artesano" [The systematization of artisan knowledge], en *La educación técnica en Puebla durante el Porfiriato: La eseñanza de las artes y los oficios* [Technical education in Puebla during the Porfiriato: Teaching arts and crafts], ed. María de Lourdes Herrera Feria (Puebla: Benemérita Universidad Autónoma de Puebla, Sistema de Investigación "Ignacio Zaragoza", Universidad Tecnológica de Puebla, Secretaría de Educación Pública, 2002), 51–52; Beatty, *Technology and the Search for Progress in Modern Mexico*, 193–194.

45. Verónica González Villalobos, "Una solución a la pobreza: el establecimiento de las escuelas de artes y oficios en México durante el siglo XIX. El caso jalisciense" [A solution to poverty: the establishment of arts and crafts schools in Mexico during the 19th century. The Jalisco case], *Historelo* 4, no. 8 (2012): 145–171.

46. Carlos Illades, *Hacia la república del trabajo: la organización artesanal en la ciudad de México, 1853–1876* [Toward the republic of work: The artisan organization in Mexico City, 1853–1876] (México: El Colegio de Mexico, 1996); Sonia Pérez Toledo, *Los hijos del trabajo* [The children of work] (México: El Colegio de Mexico, 2005).

47. *De la enseñanza de artes y oficios a la educación técnica, 1810–1909, La educación técnica en México desde la Independencia, 1810–2010* [From the teaching of arts and crafts to technical education, 1810–1909, Technical education in Mexico since Independence, 1810–2010] (Mexico: Instituto Politécnico Nacional, Presidencia del Decanato, 2011), 61–67.

48. Eguiarte Sakar, *Hacer ciudadanos*, 49–54.

49. *Enseñanza de artes y oficios*, 85–86.

50. To learn more about this pedagogical idea, see: Thomas Joseph Fiala, "Brougham, Hegel, and American Educational Reformers: European Middle Class Ideology and Its Relationship to American Educational Reform, 1826 to 1883" (PhD diss., University of Illinois at Urbana-Champaign, 1996), 28, www.ideals.illinois.edu/handle/2142/19351.

51. Jesús Vargas Valdés, "La utopía de la educación porfiriana en Chihuahua 1890–1910" [The utopia of Porfirian education in Chihuahua 1890–1910], *UEHS-La fragua de los tiempos* (Conferencia expuesta en Casa Siglo XIX, Chihuahua, 2010).

52. Anne Staples, *Recuento de una batalla inconclusa* [Recount of an unfinished battle] (El Colegio de Mexico, 2005), 250; 393.

53. Villalobos, "Una solución a la pobreza," 154.

54. Feria, "La sistematización del saber artesano."

55. *Enseñanza de artes y oficios*, 96–97.

56. Manuel Dublán y José María Lozano, eds., "Decreto del gobierno. Se establece una escuela de artes y oficios, 18 de abril de 1856, No. 4680" [Governmental decree. Establishment of a school of arts and trades, April 18, 1856, No. 4680], *Legislación mexicana o colección completa de las disposiciones legislativas expedidas desde la Independencia de la República* [Mexican legislation or a complete collection of legal dispositions starting with the independence of the republic], vol. 8 (Mexico: Imprenta del Comercio, 1877).

57. Dublán y Lozano, eds., "Decreto del gobierno, "159.

58. Manuel Dublán y José María Lozano, "Reglamento Expedido Para El Nombramiento de Alumnos de La Escuela de Artes y Oficios, 10 de Octubre de 1856" [Regulation issued for the appointment of students of the school of arts and crafts, October 10, 1856].

59. Manuel Dublán y José María Lozano, "Reglamento Interior de La Escuela Industrial de Artes y Oficios, 31. Julio 1857, No. 4961" [Internal regulations of the industrial school of arts and crafts, July 31, 1857, No. 4961].

60. Eguiarte Sakar, *Hacer ciudadanos*, 179; *Enseñanza de artes y oficios*, 190.

61. Dublán and Lozano, "Reglamento Interior de La Escuela," 537.

62. Fabiola Bailón Vásquez, "La Escuela Correccional de Artes y Oficios de Oaxaca, 1889–1901" [The Correctional School of Arts and Crafts of Oaxaca, 1889–1901], *Estudios de Historia Moderna y Contemporánea de México* 44 (2012): 138–173; Dublán y Lozano, "Reglamento interior de la Escuela"; Eguiarte Sakar, *Hacer ciudadanos*, 180; *Enseñanza de artes y oficios*; Villalobos, "Una solución a la pobreza," 159.

63. María de los Ángeles Rodríguez Álvarez, "La educación técnica de la mujer en México" [The technical education of women in Mexico], *Géneros* 10, no. 29 (2003): 33–35.

64. Carmen Ramos-Escandón, *Industrialización, género y trabajo femenino en el sector textil mexicano: El obraje, la fábrica y la compañía industrial* [Industrialization, gender and female work in the Mexican textile sector: The workshop, the factory and the industrial company] (Mexico: Publicaciones de la Casa Chata, Centro de Investigaciones y Estudios Superiores en Antropología Social, 2005).

65. Ávila Galinzoga and Calvillo Velasco, *Enseñanza de artes y oficios*, 274.

66. María de la Paz Ramos Lara, "Incidencia negative en la formación de los ingenieros electricistas mexicanos en el siglo XIX por parte del sector industrial eléctrico eminentemente extranjero" [Negative impact on the training of Mexican electrical engineers in the 19th century by the eminently foreign electrical industrial sector], *La electricidad y la transformación de la vida urbana y social*, ed. Horacio Capel y Miriam Zaar (Barcelona: Universidad de Barcelona, 2019), 669–680.

67. Feria, "La sistematización del saber artesano," 54–56, 78.

68. Carlos Ortega y Juan José Saldaña, "Un anhelo de modernidad: Félix F. Palavicini y la educación técnica en México (1906–1917)" [A longing for modernity: Félix F. Palavicini and technical education in Mexico (1906–1917)], *Memorias del X Congreso Mexicano de Historia de la Ciencia y de la Tecnología* (Mexico: Sociedad Mexicana de Historia de la Ciencia y de la Tecnología, 2006), 659–672.

69. Kuntz Ficker, "De las reformas liberales a la Gran Depresión," 344.

70. Lina del Castillo, *Crafting a Republic for the World: Scientific, Geographic, and Historiographic Inventions of Colombia* (Lincon: University of Nebraska Press, 2018), 20–25, 198–199, 234–306.

71. Alberto Mayor Mora et al., *Las escuelas de artes y oficios en Colombia (1860–1960)* [The schools of arts and crafts in Colombia (1860–1960)] (Bogotá: Pontificia Universidad Javeriana, 2014).

72. Ismaria Zapata Hoyos's Master's thesis is an insightful study of the Colombian school. She also suggests a comparative study of Schools of Arts and Trades from a global perspective. Ismaria Zapata Hoyos, "Las escuelas de artes y oficios. Una forma de munidalizar el trabajo técnico y disciplinar los artesanos" [Schools of arts and crafts. A way to globalize technical work and discipline artisans] (master's thesis, Universidad EAFIT, 2016), 22–26.

73. Safford, *The Ideal of the Practical*, 205.

74. Clara Helena Sánchez, "Matemáticas en Colombia en el siglo XIX" [Mathematics in Colombia in the 19th century], *Llull* 22 (1999): 687–705; Clara Helena Sánchez, "Anotaciones para la historia de las matemáticas en Antioquia" [Annotations for the history of mathematics in Antioquia], *Lecturas Matemáticas* 26 (2005): 91–105.

75. Álvarez Olivares, "La Escuela de Artes y Oficios de Medellín," 105–106.

76. Gabriel Poveda Ramos, "Ingenieros inmigrates y mineria en la Antioquia del Siglo XIX" [Immigrant engineers and mining in 19th century Antioquia], *Tecnología y Desarrollo* 22, no. 1–2 (1987): 105–118.

77. Álvarez Olivares, "La Escuela de Artes y Oficios de Medellín," 107–113.

78. "La Escuela de Artes y Oficios del Estado Soberano de Antioquia, tiene por objeto formar artesanos instruidos, laboriosos y honrados, que con su conducta sirvan de ejemplo y que con sus conocimientos contribuyan al adelantamiento de la. industria. en el Estado, y á la reforma de nuestras clases trabajadoras." Dirección General de Instrucción Pública, ed., "Reglamento de la Escuela de Artes y Oficios del Estado Soberano de Antioquia" [Rules of the School of Arts and Trades of the sovereign state of Antioquia] (Medellín: Imprenta del Estado, 1870), 1.

79. María Cerdá Acebrón, "Experiencias de aprendizaje desde los márgenes institucionales" [Learning experiences from the institutional margins], *Investigación e Innovación en Arquitectura y Territorio* 6, no. 2 (2018).

80. Safford, *The Ideal of the Practical*, 99–101.

81. Sanhueza Cerda, "La gestación del Instituto Pedagógico de Santiago y la movilidad del saber germano a Chile a finales del siglo XIX."

PART II

Global Reach

CHAPTER 4

"The Dignity of Labor": Liberian Industrial Education in West Africa

Bronwen Everill

Introduction

In 1881, the *Lagos Times and Gold Coast Colony Advertiser* reported that Mr. Edward Morris, of Philadelphia, "a well-connected American gentleman, deeply interested in the welfare of the Negro Race, is establishing schools in Liberia combining book learning with instruction in field labour." The *West African Repository* clipping that the *Lagos Times* quoted reported that "Mr Thomas Roberts, a young man of the Vey tribe and member of the Lincoln University, delivered also an address on Economy which was said to be a tremendous answer to the modern scientific and perhaps former religious depreciation of the possibility of cultivating the Negro Race."[1] Why was a British colonial newspaper, published by the Sierra Leonean-Nigerian businessman Richard Beale Blaize, interested in the exchange of educational ideas between Liberia and the United States?

Blaize had partnered, in 1891, with the Liberian John Payne Jackson to start the *Lagos Weekly Times*, a newspaper that advocated education, self-government, and African improvement. Industrial education was a regular feature in his newspapers and political activism, and, at his death, he founded the Blaize Memorial Institute, an industrial school that operated until the 1970s.[2] Blaize's personal history as the son of Yoruba captives who had been resettled in Sierra Leone by the British antislavery naval squadron sheds some light on the function of industrial schooling in Western Africa in the late nineteenth and early twentieth centuries. Blaize had a transnational life in West Africa. He decided as a teenager to move to Lagos in 1862, shortly after its formal annexation as a British colony. Blaize set himself up in printing and ran the *Lagos Times and Gold Coast Colony Advertiser* between 1880 and 1884, before his new venture with his Liberian partner took off.

John Payne Jackson, his Liberian partner, was born in Liberia in 1848, the year of its independence from the American Colonization Society. Jackson was educated at a Methodist mission school before trying his hand at the palm oil trade in Lagos. When he was forced out of business by European competition, he decided to go into publishing, writing for Blaize's newspaper and managing it. Inspired by the pan-Africanist thinker Edward Wilmot Blyden, Jackson advocated for an end to missionary education, which led to "the paralysation and complete emasculation which afflict the native under the spell of his so-called civilized life."[3]

Jackson and Blaize were interested in Liberian-American plans for industrial education as part of a wider interest among the Atlantic diaspora of educated Africans in Liberian, Americo-Liberian, and Black American contributions to the industrial school movement's global takeoff in the late nineteenth and early twentieth centuries. The cooperation between the Liberian Jackson and the Sierra Leonean-Nigerian Blaize helps to illustrate the ways in which Liberia became a source of wider Pan-African ideas about education, and a center for experimentation.

Long before the advent of the transnational experts of the new development empire model in the postwar twentieth century, Liberians were part of a mobile class of African engineers, missionaries, and educators spreading American values and ideas on the continent. While Liberia's education system has been of interest to historians since the early twentieth century, its function within wider imperial forms of industrial and agricultural education has sometimes been overlooked because of America's unusual relationship with the colony. Rather than fitting a classic colonial definition, Liberia is somewhat of an anomaly. The rise of global and transnational histories has seen Liberia entering discussions about the limitations of imperial ambitions, the practices of colonialism beyond formal empire, and the complicated nature of Black settler colonialism in a world increasingly defined by empire, race, and capitalist expansion.[4]

This chapter explores the industrial education plans pursued in and beyond Liberia in the nineteenth century. Liberians' participation in these projects demonstrates the transnational imperial collaborations that were essential to facilitating colonial capitalism's reach in Africa, and the ways that educational expertise could be used to both reinforce and challenge racial hierarchy in African contexts. Education policy was at the heart of Liberian assertions of sovereignty in an imperial world order.

In considering Liberia, its unique situation offers a chance to reflect on the role of forms of education and arguments about education in promoting certain kinds of ideas about civilization, state-making, measures of "success" in a capitalist world system, and the role of race in those processes.[5] In particular, the establishment of the Phelps Stokes–funded Booker Washington Institute, and its relationship to the Firestone Rubber Corporation in Liberia, has generated some attention to the question of industrial education, concessionary capitalism, and mid-twentieth-century challenges to Liberia's sovereignty.[6] The fact that Liberians both actively promoted industrial education, *and* that the majority of immigrant-descended Liberians thought that industrial education was for someone else, highlights the ways that racial uplift discourses operated in a global setting.

Liberians used education policy as a way of navigating and benefiting from imperial ideas about civilization, industriousness, and Black labor. As historians Cassandra Mark-Thiesen and Moritz Mihatsch argue, "in making claims to sovereignty, the West African state of Liberia had to prove that it shared the same ideological outlook as the family of 'civilised' states,"[7] a process mirrored in other nations like Japan, Turkey, Thailand, and Ethiopia, holding on to their independence as the world was carved into imperial territories. But even before the challenges of the Firestone period, education policy for Liberians was a means of establishing and ensuring their independence and continued self-rule in a world of empires, by working in cooperation with other empires and corporations to access capital, while promoting their own state as a model of the liberal possibilities of racial capitalism. In the period leading up to the First World War, Liberians tried repeatedly to assert their own independence through changing approaches to education, both at home and abroad. Industrial education offered a way to demonstrate Liberia's participation in the world of civilized states, but the training offered by industrial education, it was argued, also offered a practical way of asserting both individual and national sovereignty. Industrial education among Liberians was not simply a colonial project or a reaction against colonialism: It was a way that Liberians could navigate questions of authenticity, Pan-Africanism, sovereignty, and racial representation within the constraints and contradictions of colonial capitalism.

Historians and Liberia

Liberia was established in 1822 by the American Colonization Society, a white-run philanthropic organization that sponsored the emigration of free and manumitted Black Americans. Black American settlers who had emigrated to West Africa settled along the coast between the Gallinas and Cape Palmas in coastal towns established with funds from U.S. state-based colonization societies. From there, they launched missionary and educational enterprises to the surrounding Dei, Gola, Kpelle, Kru, and Vai communities. But they also went further afield, assisting British colonial educational projects. Liberians were involved in promoting Christian education and agricultural and industrial education on the infamous Niger Expedition at the beginning of the 1840s. As the nineteenth century came to a close, Liberian engineers trained African workers for plantation labor, for domestic housework, and for skilled and unskilled industrial labor in places as far away as German Togo, British Nigeria, and the Belgian Congo.

From the beginning, the settlement was a source of consternation among British colonial governors in the region, who feared that an American settlement would compete with their own colonies in Sierra Leone, Gambia, and the Gold Coast and would stymie the anti-slave trade naval patrols that the British were undertaking. The British also worried that the settlers would require rescuing from themselves.[8] Recent historiography has highlighted the complex relationship between Liberia and the British Empire.[9] Across the nineteenth century, Liberia found itself alternately cooperating with British imperial projects, like the Niger Expedition and the settlement of enslaved people from captured slaving ships, and feeling coerced by the larger power to the extent of seeking formal sovereignty from the toothless American Colonization Society in order to be able to assert itself in the face of British threats to trade and territory.[10]

The colony was a hub of transnational activity and American missionary and educational influence. As the relationship with the British Empire grew increasingly complex after Liberia's independence from the American Colonization Society in 1848, the new republic also became a center for education, missionary endeavor, trade, agriculture, and industry in the region. Liberian traders and missionaries in particular were participants in wider West African and American networks that saw them traveling in the regions away from the coast, but also to Lagos, Freetown, and farther afield.[11] Richard Blaize and John

Payne Jackson's partnership was representative of a wider trend of migration around Anglophone West Africa. Black American and Caribbean travelers also came to Liberia without the intention to settle there, but as a stop on tours of West Africa, or on the way to other mission fields in Mende country, in the Cameroons, or up the Congo River.[12] The American Board of Commissioners for Foreign Missions operated outside of Liberian territory but made stops in Monrovia on their way. American Baptist missionaries traveled to mission stations in Cameroon through Liberia.[13] Reverend William Sheppard, the Black American missionary who exposed the atrocities of King Leopold's Congo, spent time in Liberia before departing for his intended mission station.

But with the establishment of Liberia College in 1850, a new group of intellectuals emerged, contributing to the articulation of theories of Black identity, progress, and relationships with the imperial projects that Liberia was simultaneously promoting and resisting. Edward Wilmot Blyden (1832–1912) embodied this transnational position, traveling from the Danish West Indies to the U.S. to Liberia, and from there spending significant amounts of time in Lagos and Freetown. His evolving views on education, racial emancipation, capitalism, and empire were both emerging from changing attitudes about racial nationalism in the later nineteenth century, and responsible for the change.[14]

Industrial and Agricultural Training

Both new emigrants to Liberia and the Vey, Gola, Kpelle, Kru, and other groups already present in the territory that would become Liberia had plenty of expertise in agriculture and industry. Iron ore was prevalent in the region, and blacksmithing accompanied cotton growing and manufacturing as local industries. A census of the colony in 1843 records brickmakers, shoemakers, carpenters, caulkers, coopers, stonemasons, seamstresses, sawyers, millwrights, tailors, weavers, shingle makers, blacksmiths, and many farmers among those who had emigrated to the colony.

In 1841, the governor of Liberia, Thomas Buchanan, established a Public Farm on Bushrod Island to experiment with sugar production. He hired an expert in tropical sugar cultivation to work with Liberian farmers to establish the methods and growing season in the new climate.[15] By 1860, Liberia had eight sugar manufacturers, and the emigrant John B. Jordan operated a steam sugarcane mill at his farm from 1856 until his death in 1862.[16]

British interests in West Africa recognized the agricultural and industrial skills available in Liberia. The British abolition leader and reformer Thomas Fowell Buxton looked to the example of Liberia in setting out the plan for his Niger Expedition and model farm in 1841.[17] The expedition stopped in Monrovia on its way to the Niger, and William Allen's diary of the expedition records that "Mr Carr secured the services of two volunteers for the model farm" from Liberia: "one of these, Ralph Moore, had been accustomed to the management of cotton plantations on the banks of the Mississippi. He was an intelligent, well-conducted negro."[18] While the model farm was a failure and many on the expedition died, leading to a (temporary) British retreat from these model farm schools, the *Liberia Herald* reported: "We learn that the two young men who joined the *Albert* at this place were on board and that not one of the coloured men attached to the expedition died of fever showing that—had the first design been pursued of employing experienced merchant captains, and intelligent coloured men from Sierra Leone and the other colonies along the coast—the enterprise would not have failed from the same causes."[19] Learning the same lesson, British imperial policy favored "African" initiative for agricultural and industrial education from the mid-1840s. When Sierra Leonean and British colonists began moving to Abeokuta, in modern Nigeria, after the Niger Expedition, they brought with them a missionary-driven plan for agricultural training based on model farms. The leader of the Church Missionary Society (CMS) mission, Reverend Henry Townsend, focused on tobacco production, and drew on the expertise of Thomas Jefferson Bowen, an American Baptist missionary, and Andrew Wilhelm, an African CMS agent, to set up a model farm in Ake.[20]

As Liberia's success in agriculture took off after the U.S. Civil War, the *West African Reporter*, published in Freetown, celebrated that Allen Hooper, the largest coffee plantation owner in Liberia, "is now on a visit to this Colony, and is anxious, during his stay, to give the benefit of his experience in the cultivation of that plant or in tropical agriculture generally to any here who may be anxious to engage in such enterprises."[21] Hooper had 70,000 to 80,000 coffee plants in the mid-1850s, and, from 1865, the price of Liberian coffee rose from 14 cents to 42 cents per pound.[22] Sir Samuel Lewis, a Sierra Leonean barrister trained in England, and Sierra Leone's first knight, took up this call, purchasing 1,350 acres on which to grow "Liberian Coffee," which had been purchased from a model farm at Fourah Bay College, and which would be grown with trained agricultural laborers "imported" from Liberia.[23] The *West African Reporter* lamented the lack of agricultural training in the colony, commenting that "no community can have any permanent prosperity without a division of labor and a diffusion of industry and property."[24] Lewis's "importation" of trained Liberian agricultural specialist labor hints at one of the major reasons for Liberia's growing position in the industrial education movement. Liberia provided a model of Black independence, as well as a practical training ground for skilled laborers. Liberian maritime workers (predominantly Kru) were recruited by Elder Dempster shipping lines to work from Liverpool to the Congo, and Liberian agricultural expertise was similarly sought out by producers in Sierra Leone and Nigeria.

Liberal Versus Vocational Education

The skills that were important in establishing the new colony and attracting British attention were not, however, what many Liberian parents wanted for their children. The emigrant George Erskine wrote that "the thing most to be deplored in this colony is the want of a good school, and an enlightened teacher for poor children whose parents cannot school them; for in this tropical country, where there is an abundance of gold, ivory, cam-wood, coffee and sugar, indigo, and many lucrative fruits, we could have an enlightened community. There is nothing to hinder it from rising to a level, in point of eminence, wealth, and power, among the most refined nations of the earth."[25] Education was, from the very start, portrayed as essential to building the colony in order to demonstrate that Black people could operate and succeed in a world of empires.

Across a century of education policy, the Black Americans who moved to Liberia established a two-tiered approach to education that mirrored other colonial empires in West Africa: Missionary education focused on training lawyers, civil servants, and clerical workers was retained for an Americo-Liberian–descended elite who ran the country; and industrial and agricultural training aimed at Indigenous Liberians and other West Africans had no intention that this technical education would create social mobility and an educated middle class. This was the argument for what W. E. B. Du Bois later described as the "talented tenth" in practice, and Liberia's reputation as an independent Black republic gave it the advantage of appealing to Black leaders in West African colonies, since "African intellectuals . . . favour[ed] literary education and were likely to resist if the vocational education of the

British working class was introduced in Africa. But an American model gave a much better impression, in view of the African respect for black American leaders."[26]

The argument for the "Bible and the Plough" shaped early missionary efforts in West Africa. A combination of religious instruction and mechanical and agricultural education was set out by the Wesleyan and Church Missionary Societies that operated chiefly in Sierra Leone. But from the start, both Sierra Leone and Liberia were problematic locations for this approach. Missionaries, at least at the beginning, were most interested in using the settlements as a base from which to set out to "civilize" Indigenous Africans. But in both colonies, the governments also wanted the missionaries to provide education for the population of Black settlers. This rankled Liberians who portrayed themselves as *part* of the civilizing mission, bringing Christianity and American education to Africa, rather than needing the services of the missionaries. As one emigrant wrote in the 1840s, "I told you that I was a mechanic, and so I am in a two fold sense, that is, I am an ordained elder in the Baptist church, and have preached the Gospel these fifteen years, and . . . I am a workman and understand the brick business in all its different branches, to make, to build and to finish."[27]

Meanwhile, Liberia's own settlers saw themselves as needing more than the rudimentary missionary education on offer, and certainly not in need of evangelizing. Colonists pushed back with their own demands of missionary education—literature, French, dancing lessons, and the well-rounded education that would make them suitable for administrative posts were in high demand.[28] As people like W. E. B. Du Bois would later argue, Americo-Liberians both were capable of and *needed* the classical education that would allow them to demonstrate their ability to govern a sovereign republic and to be taken seriously by other states and empires. In Sierra Leone, "The younger generation" in the 1840s and 1850s "persistently identif[ied] civilisation with westernization and clerical and administrative jobs. Farming, as they knew it from their parents, was not thought suited for the world supposedly opening before them."[29] As historian Kehinde Olabimtan writes about Sierra Leone, "Members of the [African Improvement Society] appreciated the scale of this problem and knew that the success of their efforts lay in the future."[30] In Liberia, "The association of manual labor with slavery" was assumed to be behind the rejection of yeoman farming in favor of large-scale plantations or a form of tenant farming, trade, and politics, but the elite of Liberia were as often industrialists and large-scale farmers as they were following commercial and intellectual pursuits.[31] However, parents understandably wanted their girls trained to be middle-class Victorian wives and mothers, not domestic servants; they wanted their boys educated to be clerks and government officials, not manual laborers.

The demand for liberal arts and university education regularly featured in the messages back from Liberian settlers to the various state colonization societies that had sponsored them. The New York Colonization Society established a school "under the patronage of the N.Y. Ladies' Society for the promotion of education in Africa." The principal of the school noted that the school had received "the sum of $20 contributed in cash, work, and plank, by the parents and guardians of the scholars, for fitting up the school room."[32] The school advertised its offerings as "Spelling and Defining, Reading, Writing, Geography, 1st and 2d Book, (Goodrich;) Arithmetics, written and intellectual, (Smith and Colburn's) Grammars, History, Composition, and Declamation. Instruction in Needle Work twice a week by Mrs. James," and cost one dollar per quarter.[33] In 1851, two Boston-based philanthropic organizations began raising funds to establish the first degree-granting institution in West Africa. Their argument for the establishment of the college was that missionary education had been proceeding well in the Mende Mission, in Sierra Leone, and along the Liberian coast, and there was now sufficient student demand for a college.[34] The suggested aim of the college was to train "teachers who are not raised up in mission schools," which had a primary focus on Bible teaching and basic literacy.[35]

But the desire to establish a college in Liberia also arose from the question of sovereignty. Liberia's independence in 1848 put the new republic in a precarious global position. It had sought independence in part to enable the state to collect tariffs on trade within territory it claimed as its own, but which British and Sierra Leonean traders disputed. Liberia needed those tariffs in order to command the kind of wealth that would allow it to be recognized as a comparable "civilized" power in the region, one that could not be absorbed into another empire. But the British traders largely ignored the state's demands; the Liberian navy was donated by the French; the U.S. refused to recognize an independent Black republic before the Civil War; and the finances of the Liberian government were increasingly tied up with British loans. These factors contributed to Liberia's leadership seeking out ways to improve its international standing and capitalize on areas in which it had expertise.

Education became one of the most important ways that Liberians sought to position themselves as indispensable in the world of empires. The Trustees of Liberia College reported that the Vai, who had developed their own writing system and established their own schools, were originally thought to be within the remit of Liberian missionaries, "but after the suppression of the slave trade at Gallinas" by the British squadron in the 1830s, "more than 1,000 liberated Africans were carried thence to Sierra Leone and three sons of the chiefs were sent to the Grammar School at Freetown for their education," which had then expanded the British Church Missionary Society's interest in Vai country.[36] The report then outlined that the British, through the Church Missionary Society and increasingly through the efforts of Liberated Africans trained in Freetown, had a "line of stations commenc[ing] at Dix Cove, about 320 miles east of the eastern boundary of the Maryland Government," the final county of Liberia, "and extends about 350 miles eastward to Badagry" in Nigeria.[37] The case for a college in Liberia was framed as a point of competition with the British Empire, whose educated subjects were rapidly spreading the empire's influence along the West African coast. The report stated that "Liberia, being an independent nation, has the greatest need of a College," because "it cannot safely remain dependent on other nations for pastors and teachers, for legislators and magistrates."[38] If Liberians—either settlers or the children of groups like the Vai—had to go to colleges and universities elsewhere, "they would come back with a low esteem for the country that could not educate them" and would have British imperial interests at heart, rather than the Liberian nation's.[39]

By 1852, the committee that had established Liberia College expanded its mission to respond to demand: "We find, therefore, at the close of the year, in all the colonized portions of Western Africa, an increasing need of a College, to complete the education of the increasing numbers who will be prepared to avail themselves of its advantages."[40] Liberia College, building on the "college mania" taking place in small towns across the U.S., attracted students and teachers from around the Black Atlantic diaspora.[41] The college was intended for the sons of Americo-Liberian settlers, and was seen as "essential for the maintenance of moral and intellectual standards necessary to maintaining free, republican government," including Greek, Latin, math, history, and French.[42] At the same time, the Trustees wanted to ensure "the progress of . . . civilized labor, among the natives" to develop "new forms of productive industry," which would allow the Republic to sit amongst the likes of "Great Britain, France, and other leading nations of Christendom."[43]

Industrial Education and the Limits of the Civilizing Mission

In promoting the establishment of Liberia College, the Trustees highlighted that Liberia would attract students "from all parts of Western Africa" because "in Liberia the students will not be under the control of another race."[44] After the U.S. Civil War, and with the growth of European empires in West Africa, "racial improvement" became an important, and central, problem on which education policy focused. Industrial education was seen as a way of instilling education without the classic missionary education that would "destroy . . . race pride" and "love for our race" by teaching the student to "look down with contempt upon the customs and institutions of his own country."[45] Educated Black people sought ways of promoting self-help and the corresponding independence it offered in contrast to "slavish" wage or tenant labor and obsequiousness to the white civilizing mission, which was increasingly regarded as an inferiority trap.[46]

There was widespread concern among this rising generation of West Africans that their experiences of missionary education had alienated Black people from their "true nature," leaving them in a position of inferiority relative to white people in the imperial world; they were also dislocated from African culture, and therefore without the ability to be fully independent from imperial culture. The *Lagos Times*, for instance, published a letter from a reader in Freetown who worried that the lack of "industrial teaching" or the "teaching of civilized Agriculture" "has been one of the chief causes why a very large portion of the sons of Sierra Leone are to be found almost everywhere along the West Coast of Africa and in other places, having been compelled to leave their homes in search of employment either as Clerks or Traders."[47] The reader hoped that the government would enact an education bill that would provide "for the establishment of Institutions for training the young in civilized Agriculture and other Industrial pursuits" for the promotion of "a healthy independence."[48]

Independence was invoked at both the individual and the national level, embodying the desire both for freedom and for sovereignty.[49] Someone with appropriate vocational training in skilled labor would be able to convert those skills into an independent livelihood; a nation with enough skilled laborers would be able to produce its own food and manufactured products, rather than rely on imports. To this end, a soap manufactory was established in Liberia in 1848, "to keep in the Republic the large sum which have been annually drawed [sic] off for the article."[50] The rhetoric surrounding the reliance of Liberia on British imports in particular became so embedded in national rhetoric that, as historian William Allen noted, it came to define the historical view of Liberia's elite in the nineteenth century.

Historian Juliet Walker points to the existence of the business relations committee of the National Council of Colored People to argue that "expressions of black entrepreneurship were shaped by the black community, both slave and free, for the promotion of black economic interests rather than for the profits of slave-owners or white employers,"[51] and, across the nineteenth century, the political ideology underlying that argument for Black economic nationalism was attached to political pressure to demonstrate independence and ability. For instance, the Black American abolitionist Martin Delany's writings for the *North Star* in the late 1840s emphasized the importance of "Practical Industry, and money-making business enterprises." He argued that "we have been heretofore taught that these things were unfit for us, as they interfered with our prospects of heaven. This our oppressors taught us to prevent us from competing with them in business."[52]

As ideas of racial nationalism began to replace the "civilizing mission" in imperial discourses, people like Edward Wilmot Blyden, whose career was launched at Liberia College, began to argue in the 1870s that a loss of connection to their African roots had alienated Black people from their authentically African selves and thereby diminished them. While Blyden approached this question from the perspective of Black uplift, it was a more general one on the minds of educational experts more broadly, and one which preoccupied the popular writers of empire, who had a growing concern about authenticity and purity. Masculinity was seen to be in crisis across the world, and industrial and agricultural training in part gained popularity among the leaders in West Africa because they were themselves worried about their dislocation from the traditional masculinity that had been displaced by feminized civilization. While almost never explicitly addressed in these writings, women's practical education was largely focused on training for domestic service, something that largely divorced it from the political meaning and import of the men's industrial schooling model. Writers such as Richard Burton and Mary Kingsley added to the anxiety among West African elites by mocking their "civilized" habits and promoting the handover of colonial power to "authentic" chiefs, uncorrupted by the "mimicry" of European manners and customs.[53]

Blyden was concerned that educational opportunities must exist *in* Africa in order to allow for the development of systems of education that did not "de-Africanize" while they "civilized." He wrote that "all the expense we are going to, in educating our children abroad, will end in failure, disappointment and reaction; simply because such education must necessarily divert them from their natural bent, or confuse the clearness of their native instincts."[54] The Nigerian nationalist Herbert Macaulay, who worked closely with John Payne Jackson in the early 1910s, argued that "our fathers were taught that everything European is superior, and everything Native is degrading, superstitious, and immoral . . . If we want Lagos, Benin, Abeokuta, Ibadan, and Yorubaland to be included among the leading countries of the world, we must first of all identify ourselves with our Country."[55]

At the same time, European empires found themselves in a bind in their African colonies. Having premised their expansion at least in part on the promises of the Brussels Act to end the slave trade and slavery, and with the negative attention drawn to examples of enslaved and coerced labor in the Belgian Congo and São Tomé and Principe, colonial officials were always on the lookout for new ways of commanding African labor for as little cost as possible. There was a balance to be struck in education between technical advancement and the reduction of overall poverty, and social advancement and the reduction of inequality. While development was a goal, the British Empire Advisory Committee on Education summarized the underlying idea: "There is obviously an intimate connection between educational policy and the economic development of a territory. Educational policy must be planned with reference to the kind of life the pupils may be expected to lead."[56] That is to say, certain jobs, lifestyles, and social worlds were never intended for the African subjects of the empire; for them, education was training and development, but not social advancement.

As colonial empires sought to solve the "labor problem" in Africa, they looked to the U.S. Southern states for inspiration.[57] Sociologists, anthropologists, and economists in the United States participated in diagnosing and prescribing remedies for the "Native Problem."[58] Zimmerman has written about the German use of the Tuskegee system in German Togo, for instance.[59] Graduates of Tuskegee and Hampton migrated to employment as managers and teachers in Nigeria, Congo, South Africa, and British East Africa.[60]

Liberians were as interested in the new industrial education model as the European empires, and, because of the perceived shared racial characteristics of Africans and Black Americans, Liberia offered a point of reference for the founders of the main U.S. industrial schools as well. Hampton Institute, founded in Virginia in 1868 by the northern philanthropist Samuel

Armstrong, looked to "the habit of labour" to save formerly enslaved people from ruin.[61] The focus was on the "dignity of labour." Hampton and Tuskegee both aimed to teach "the 'right' ideas of citizenship, the duties of labourers, and the history of race development."[62] Armstrong was a proponent of Liberian emigration, and addressed the American Colonization Society in 1879 in his capacity as principal of the Hampton Institute. His speech noted that "Africa—Liberia as one of its open doors—is the field for an Anglo-African crusade." However, he also noted that as in Jamaica, "where the whites, colored and blacks are completely severed socially," an informant had told him that "it appears in Liberian politics" as well, and that emigration would not entirely remove the operation of "prejudice of color."[63] Neither industrial education nor emigration to Liberia were intended to erase racial prejudice—they were intended to help people fulfil their specific, natural place in the global order.

Blyden toured Tuskegee and declared that he would implement a similar program as president of Liberia College. In his argument for a liberal education for Africans, he wrote that Liberia College should be relocated to the interior of the country, "where the students may devote a portion of their time to manual labor in the cultivation of the fertile lands which will be accessible, and thus assist in procuring the means from the soil for meeting a large part of the necessary expenses; and where access to the institution will be convenient to the aborigines." He went on to worry that "the people generally are not yet prepared to understand their own interest in the great work to be done for themselves and their children."[64] However, this was contrasted, for instance, in Vai attitudes toward education in the Liberian interior—they were promoting their own schools and language training, and they maintained "traditional concepts of 'proper' and 'improper' types of work," even as they moved to urban areas, focusing on trades like goldsmithing and cloth manufacturing.[65] This suggests that even Blyden was not entirely aware of the existing ideas of education that were circulating among Indigenous Liberians.

Tuskegeeism became the new watchword in Black education both in the U.S. and in the European colonial empires in Africa. German colonists' use of Tuskegee students in Togo led them "to introduce into all the schools of that colony a system for the training of boys in handwork."[66] The focus on "hard work, industry and thrift" was intended to focus on economic, rather than political, power.[67] After 1915, "the emphasis on self-reliance and racial advancement was replaced by an appeal for blacks to aid in the exploitation of natural resources and the development of the nation . . . creating a black work force that would willingly respond to the demands of industry."[68] But it was always a matter of increasing the productive labor force, rather than educating people "above their station": "If the white southerners had to permit the Negro to obtain any education at all, they wanted it to be of the sort that would make him a better servant and labourer, not that which would train him to rise out of his 'place.'"[69] There was a civilizational and developmentalist agenda at work, but one still fundamentally based on hierarchies and paternalism.

As European colonial empires engaged with these American plans, African entrepreneurs such as Richard Blaize, the Nigerian *Lagos Times* newspaper founder, translated their own success into philanthropy that mirrored institutions like the Phelps Stokes fund. Blaize died shortly after accompanying Oba Gbadebo Alake to England, leaving £3,000 in his estate for the free training of boys in technical work in Abeokuta. This gave birth to the Abeokuta Industrial Institute, which opened on February 14, 1908. The institute combined training with labor: Every apprentice was expected to serve a period of four years, at the end of which he was given a set of tools worth £6. The first manager of the institute was a European, Horace J. Barwick. In the early days, the Nigerian Oriade Solotan also took charge of the institute when a European manager was not available.[70]

Not wanting to miss out on these opportunities, and the possibilities offered by these funds, Liberia's own educated elite men seized on their unique position as Americans in Africa.[71] Liberians migrated around the British West African Empire, as well as to the United States, both as elites bringing their expertise, and as workers bringing their skills to an expanding concessionary colonialism in need of labor. Plenyono Gbe Wolo, who was the son of a Kru paramount chief, is one example. As historian Gregg Mitman recounts, in the 1910s, Wolo attended Harvard and Columbia, as well as a summer at Hampton Institute, and was sponsored by the Phelps Stokes Fund to establish a school in Grand Cess that would focus on teaching "practical techniques of raising foodstuffs and the introduction of native crafts" based on the industrial model.[72]

If the United States offered a solution to the labor problem through industrial education, then Liberians could demonstrate the American principles in Liberia and beyond in order to legitimate their role as a sovereign power in Africa. Blyden argued that Liberians were uniquely placed to educate people in Africa because "residence in America has conferred upon him numerous advantages . . . it has predisposed him in favour of civilisation."[73] Despite disagreements between various branches of Black American ideas about the race question, W. E. B. Du Bois, Marcus Garvey, and Booker T. Washington all found themselves on the same side in closing ranks with Liberia's leaders in order to promote the Black Republic as a sovereign international power that demonstrated the potential for African leadership.[74]

In Liberia, leaders of the movement "saw industrial education as a method of erasing poverty, illiteracy and ill-health among Liberia's indigenous population."[75] The Liberian government could show itself to be developing the population of the country and improving the living standards and employability of its labor force, in line with similar efforts by the French and British Empires to act as managers of the "development" process of the Indigenous population. But unlike the French and British, advocates of Liberian models of industrial education, such as Jackson and Blaize, emphasized the industrial model as being the opposite of European "civilizing" missionary education. Jackson, in particular, wrote about the civilizing mission as "devilization" rather than "civilization," and argued that a focus on practical training would remasculinize the African population through an assertion of their self-sufficiency, and would undermine the kinds of submissive relationships that developed in European missionary education.[76] Industrial, rather than missionary, education would provide the means for Black people around the diaspora to assert their independence, and the ability to provide industrial education to its Indigenous population would, ironically, enable Liberia itself to demonstrate its standing among the civilized nations of the world.

Conclusion

As Liberians saw their borders being challenged by imperial expansion, and as their government relied on British loans to ensure they could "effectively occupy" their territory after the Scramble for Africa in the 1880s and 1890s, independence was everything. For people like John Payne Jackson, a Liberian emigrant to Lagos, who inspired the Sierra Leonean-Nigerian Richard Blaize to found an industrial school, Liberia's place in the world was dictated by its relationship with the British Empire in West Africa, and by its connections to the African Diaspora in the U.S. Promoting vocational training was a way of connecting Liberia's mission to the mission of the British Empire, ensuring that the dominant power in the region was invested in Liberia's success. But it also provided a way for Liberians to assert their sovereignty and their relevance to the cause of Pan-African nationalism. Industrial education was a vision of Pan-African influence and independence; it was celebrated in West African newspapers and championed by leaders like Edward Wilmot Blyden, even while it played into imperial designs.

Donald Spivey has argued that the ideal of industrial education "was white rule and the method was a dissemination of educational ideas that were conducive to subordination and exploitation."[77] For Liberia, it was much more complex. In the period before the First World War, industrial education was both about a perception of Liberia's place in the world and an opportunity to ensure a certain position as that world changed around it. It was about the double consciousness of needing to represent "Black potential" to act as an equal among "civilized" nations, while also addressing the identity crises of the late nineteenth century. It was about using whatever means it had at its disposal to ensure that Liberia was not subsumed into the British (or, less likely, the French) Empire, in part by ensuring that Liberia's elites had effective control over and domination of the Indigenous populations of Liberia. Education became a means of playing the imperial game.

Notes

1. *Lagos Times and Gold Coast Colony Advertiser*, September 14, 1881, 3.

2. Nozomi Sawada, "The Educated Elite and Associational Life in Early Lagos Newspapers: In Search of Unity for the Progress of Society" (PhD diss., University of Birmingham, 2011); Tom Forrest, *The Advance of African Capital: The Growth of Nigerian Private Enterprise* (Charlottesville: University Press of Virginia, 1994); Ajayi Kọlawọlẹ Ajisafe, *History of Abeokuta* (Abeokuta: Ola Fola Bookshop, 1998), 174.

3. Cited in Fred I. A. Omu, "Journalism and the Rise of Nigerian Nationalism: John Payne Jackson, 1848–1915," *Journal of the Historical Society of Nigeria* 7, no. 3 (1974): 525.

4. Caree A. Banton, *More Auspicious Shores: Barbadian Migration to Liberia, Blackness, and the Making of an African Republic* (Cambridge: Cambridge University Press, 2019); Cassandra Mark-Thiesen and Moritz A. Mihatsch, "Liberia an(d) Empire?: Sovereignty, 'Civilisation' and Commerce in Nineteenth-Century West Africa," *Journal of Imperial and Commonwealth History* 47, no. 5 (2019): 884–911; A. Zimmerman, *Alabama in Africa: Booker T. Washington, the German Empire and the Globalization of the New South* (Princeton: Princeton University Press, 2010); Lisa A. Lindsay, *Atlantic Bonds: A Nineteenth-Century Odyssey from America to Africa* (Chapel Hill: University of North Carolina Press, 2016); Bronwen Everill, *Abolition and Empire in Sierra Leone and Liberia* (Basingstoke: Palgrave Macmillan, 2013); Beverly C. Tomek and Matthew J. Hetrick, eds., *New Directions in the Study of African American Recolonization* (Gainesville: University Press of Florida, 2017); Sebastian N. Page, *Black Resettlement and the American Civil War* (Cambridge: Cambridge University Press, 2020).

5. See Olúfẹ́mi O. Táíwò, *Elite Capture: How the Powerful Took Over Identity Politics (and Everything Else)* (London: Pluto Press, 2022).

6. Louise Johnston, "Tuskegee in Liberia: The Politics of Industrial Education, 1927–1935," *Liberian Studies Journal* 9, no. 2 (1980–1981): 61–68; Leigh A. Gardner, *Sovereignty Without Power: Liberia in the Age of Empires* (Cambridge: Cambridge University Press, 2022); Gregg Mitman, *Empire of Rubber: Firestone's Scramble for Land and Power in Liberia* (New York: New Press, 2021).

7. Mark-Thiesen and Mihatsch, "Liberia an(d) Empire?" 891.

8. Everill, *Abolition and Empire*; Padraic X. Scanlan, *Freedom's Debtors: British Antislavery in Sierra Leone in the Age of Revolution* (New Haven: Yale University Press, 2018); Lamin Sanneh, *Abolitionists Abroad: American Blacks and the Making of Modern West Africa* (Cambridge: Harvard University Press, 1999).

9. Christine Whyte, "A State of Underdevelopment: Sovereignty, Nation-Building and Labor in Liberia, 1898–1961," *International Labor and Working-Class History* 92 (2017): 24–46; Christine Whyte, "Between Empire and Colony: American Imperialism and Pan-African Colonialism in Liberia, 1810–2003," *National Identities* 18, no.1 (2016): 71–88.

10. Anthony Niblette Wojloh, "Boundary Disputes Between Liberia and Colonial Powers, 1911–1923" (PhD diss., Howard University, 1971); Eugene S. Van Sickle, "Reluctant Imperialists: The U.S. Navy and Liberia, 1819–1845," *Journal of the Early Republic* 31, no. 1 (2011): 107–134.

11. Lindsay, *Atlantic Bonds*; Banton, *More Auspicious Shores*; Kristin Mann, "Marriage Choices Among the Educated African Elite in Lagos Colony, 1880–1915," *International Journal of African Historical Studies* 14, no. 2 (1981): 205; Robert Murray, "Bodies in Motion: Liberian Settlers, Medicine, and Mobility in the Atlantic World," *Journal of the Early Republic* 39, no. 4 (2019): 615–646.

12. James Campbell, *Middle Passages: African American Journeys to Africa, 1787–2005* (London: Penguin Books, 2006); Bronwen Everill, "Bridgeheads of Empire? Liberated African Missionaries in West Africa," *Journal of Imperial and Commonwealth History* 40, no. 5 (2012): 789–805; James Fairhead, Time Geysbeek, Svend E. Holsoe, and, Melissa Leach, *African-American Exploration in West Africa: Four Nineteenth-Century Diaries* (Bloomington: Indiana University Press, 2003).

13. Charles W. Weber, *International Influences and Baptist Mission in West Cameroon: German-American Missionary Endeavor Under International Mandate and British Colonialism* (New York: Brill, 1993).

14. Teshale Tibebu, *Edward Wilmot Blyden and the Racial Nationalist Imagination* (Rochester: University of Rochester Press, 2012).

15. William Ezra Allen, "Historical Methodology and Writing the Liberian Past: The Case of Agriculture in the Nineteenth Century," *History in Africa* 32 (2005): 31–32.

16. Allen, "Historical Methodology and Writing the Liberian Past," 26.

17. Thomas Fowell Buxton, *The African Slave Trade and Its Remedy* (London, 1839).

18. William Allen and Thomas Richard Heywood Thomson, *A Narrative of the Expedition Sent by Her Majesty's Government to the River Niger, in 1841, Under the Command of Captain H.D. Trotter, R.N.* (London: J. Murray, 1848), 89.

19. *Liberia Herald* XI, no. 12, October 31, 1842.

20. Kehinde Olabimtan, "Church Missionary Society Projects of Agricultural Improvement in Nineteenth-Century Sierra Leone and Yorubaland," in *Commercial Agriculture, the Slave Trade and Slavery in Atlantic Africa*, ed. Robin Law, Suzanne Schwarz, and Silke Strickrodt (Woodbridge: James Currey, 2013), 217.

21. "The People of Sierra Leone—Their Progress, and Their Needs," *West African Reporter*, April 23, 1879, 2.

22. William Ezra Allen, "Sugar and Coffee: A History of Settler Agriculture in Nineteenth-Century Liberia" (PhD diss., Florida International University, 2002), 130–133.

23. "Agriculture in Sierra Leone," *Methodist Herald and West African Educational Times*, October 8, 1884, 2.

24. "The People of Sierra Leone—Their Progress, and Their Needs," *West African Reporter*, April 23, 1879, 2.

25. Letter from George Erskine to R.R. Gurley, March 9, 1833, in *African Colonization*, Historical Society of Pennsylvania Tw*.3543, 41.

26. Shoko Yamada, "Educational Borrowing as Negotiation: Re-Examining the Influence of the American Black Industrial Education Model on British Colonial Education in Africa," *Comparative Education* 44, no. 1 (2008): 29.

27. The American Colonization Society, *The African Repository and Colonial Journal* 19, no. 3 (1843): 93.

28. Silke Strickrodt, "African Girls' Samplers from Mission Schools in Sierra Leone (1820s to 1840s)," *History in Africa* 37 (2010): 189–245.

29. Olabimtan, "Church Missionary Society," 211.

30. Olabimtan, "Church Missionary Society," 211.

31. M. B. Akpan, "Black Imperialism: Americo-Liberian Rule over the African Peoples of Liberia, 1841–1964," *Canadian Journal of African Studies/Revue canadienne des Études africaines* 7, no. 2 (1973): 217–236.

32. *Liberia Herald* XV, no. 11, March 19, 1847.

33. *Liberia Herald* XV, no. 11, March 19, 1847.

34. *First Annual Report of the Trustees of Donations for Education in Liberia* (Boston, 1851), 14–15.

35. *First Annual Report*, 14–15.

36. *First Annual Report*, 17.

37. *First Annual Report*, 17.

38. *First Annual Report*, 17.

39. *First Annual Report*, 21.

40. *Second Annual Report of the Trustees of Donations for Education in Liberia* (Boston, 1852), 12.

41. Thomas Livingston, "The Exportation of American Higher Education to West Africa: Liberia College, 1850–1900," *Journal of Negro Education* 45, no. 3 (1976): 251.

42. Livingston, "The Exportation of American Higher Education to West Africa," 254–255

43. *First Annual Report*, 23.

44. *First Annual Report*, 24.

45. As quoted in the *Lagos Standard*, May 3, 1904, 2.

46. Akpan, "Black Imperialism," 217–236.

47. Letter from "A Wellwisher of His Country," *Lagos Times and Gold Coast Colony Advertiser*, November 23, 1881, 3.

48. Letter from "A Wellwisher of His Country," 3.

49. David Kazanjian, *The Brink of Freedom: Improvising Life in the Nineteenth-Century Atlantic World* (Raleigh: Duke University Press, 2016).

50. *Liberia Herald*, November 23, 1848.

51. Juliet E. K. Walker, "Racism, Slavery, and Free Enterprise: Black Entrepreneurship in the United States Before the Civil War," *Business History Review* 60, no. 3 (1986): 375.

52. Martin R. Delany, *North Star*, March 16, 1849, 2; Martin R. Delany, *North Star*, April 20, 1849, 2; Robert S. Levine, ed., *Martin R. Delany: A Documentary Reader* (Chapel Hill: University of North Carolina Press, 2003).

53. Richard F. Burton, *Wanderings in West Africa* (London, 1863); Mary Kingsley, *Travels in West Africa* (London, 1897); Vivian Bickford-Smith, "The Betrayal of the Creole Elites, 1880–1920," in *Black Experience and the Empire*, ed. Philip D. Morgan and Sean Hawkins (Oxford: Oxford University Press, 2004), 194–227; Catherine Hall, *Civilising Subjects: Metropole and Colony in the English Imagination, 1830–1867* (Chicago: University of Chicago Press, 2002).

54. Edward W. Blyden to William Grant, member of the Legislative Council of Sierra Leone, in *The Negro*, June 5, 1872.

55. *Lagos Standard*, October 26, 1910, 5.

56. Extracts from Memorandum of British Empire Advisory Committee on Education, *International Education Board Collection*, File No. 221, 1935, 2; cited in Donald Spivey, "The African Crusade for Black Industrial Schooling," *Journal of Negro History* 63, no. 1 (1978): 7.

57. Frederick Cooper, *Decolonization and African Society: The Labor Question in French and British Africa* (Cambridge: Cambridge University Press, 1996); I. K. Sundiata, "Prelude to Scandal: Liberia and Fernando Po, 1880–1930," *Journal of African History* 15, no. 1 (1974): 102.

58. Raymond Leslie Buell, *The Native Problem in Africa* (New York: The Macmillan Company, 1928).

59. Zimmerman, *Alabama in Africa*; Sven Beckert, *Empire of Cotton: A Global History* (New York: Allen Lane, 2016).

60. Spivey, "The African Crusade," 2.

61. Yamada, "Educational Borrowing as Negotiation," 26.

62. Thomas Jesse Jones, "Social Studies in the Hampton Curriculum," *Southern Workman*, 1906, 11–13.

63. Samuel Armstrong, *Emigration to Liberia: An Address Delivered Before the American Colonization Society, January 21, 1879* (Washington: American Colonization Society, 1879), 5–6.

64. Edward Wilmot Blyden, *The Aims and Methods of a Liberal Education for Africans, Inaugural Address, January 5, 1881* (Cambridge, 1882), 11.

65. Svend Holsoe, "Vai Occupational Continuities: Traditional to Modern," *Liberian Studies Journal* 10, no. 2 (1984): 12–23.

66. Booker T. Washington, *Working with the Hands* (New York: Doubleday,1904), cited in Spivey, "African Crusade," 2; Zimmerman, *Alabama in Africa*.

67. Louise Johnston, "Tuskegee in Liberia: The Politics of Industrial Education, 1927–1935," *Liberian Studies Journal* 9, no. 2 (1980–1981): 65.

68. Johnston, "Tuskegee in Liberia," 65.

69. Raymond Blaine Fosdick, *Adventure in Giving: The Story of the General Education Board, a Foundation Established* (New York: Harper and Row, 1962), 85.

70. Many thanks to Lanre Davis for this information; Ajișafẹ, *History of Abeokuta*, 174.

71. Mark-Thiesen and Mihatsch, "Liberia an(d) Empire?" 891; S. Jabaru Carlon, "Black Civilization and the Problem of Indigenous Education in Africa: The Liberian Experience," *Présence Africaine* 3, no. 95 (1975): 253–268.

72. Mitman, *Empire of Rubber*, 93–94.

73. Edward W. Blyden, *Christianity, Islam and the Negro Race*, cited in Tibebu, *Edward Wilmot Blyden*, 77.

74. Tamba E. M'Bayo, "W.E.B. Du Bois, Marcus Garvey and Pan-Africanism in Liberia, 1919–1924," *Historian* 66, no. 1 (2004): 19–44; Mitman, *Empire of Rubber.*

75. Johnston, "Tuskegee in Liberia," 65.

76. Omu, "Journalism and the Rise of Nigerian Nationalism," 526; *Lagos Weekly Record*; "Devilization in Africa," *Dakota County Herald*, May 22, 1908.

77. Spivey, "African Crusade," 14.

CHAPTER 5

The Making of Christian Industrial Labor: Missionaries, Industrial Schooling, and the British Empire in India

Arun Kumar

Introduction

Industrial schools emerged as innovative institutions of nineteenth-century modernity, aiming to train economically and racially marginalized groups for an industrializing world. As other chapters in this volume illustrate, these schools targeted the poor, vagrants, and orphans in European nations, ex-enslaved Blacks and Indigenous societies in North America, and the most precarious groups in colonized societies.[1] Essentially, industrial schooling became a means of bringing the large poor population under the ambit of economic productivity and industriousness in the name of modern education that suited their life stations. A precursor of today's secular vocational and technical education, the industrial education of the nineteenth century evolved through the activities of Christian religious bodies and mission societies that were interested in civilizing "heathens" and advancing a moralistic disciplined work ethic.[2] This chapter traces the history of the late nineteenth-century Cawnpore Mission Industrial School run by the Society for the Propagation of Gospels in Foreign Parts (hereafter SPG), a Church of England missionary society, to illustrate the relationship between Christian missionaries, industrial education, and the "lowest castes" in colonial India under the British Empire. By focusing on the activities of the SPG through their Cawnpore Mission Industrial School, this chapter illustrates how Protestant missionaries in India saw manual training for natives as critical to the survival, expansion, and civilizational justification of their projects. Protestant Christian organizations did not limit themselves to providing entrepreneurial spirit and ethics in the Global North, but also involved themselves in founding and operating industrial and agricultural schools in Asia and Africa. Focusing on mission industrial schools in India allows us to explore intersections between Christianity, industrial capitalism, and native labor, and understand how ideas of Western hierarchized industrial education systems were translated into colonies marked by not only rigidities of race but also caste hierarchy. It also informs the historiography of missionary and colonial education, illustrating how colonized labor education played a key role in forming the ground-level workforce for the Indian colonial bureaucracy.[3]

Unlike the state industrial schools established in the late nineteenth century to train a skilled labor force for particular local industries/trades—glass-blowing, carpentry, and blacksmithery—or for particular employers such as railway workshops, mission industrial schools' stated objectives were to alleviate the poverty of local Christian converts and make them ready for employment.[4] Equally embedded in this project of "industrializing natives" was the pedagogy of "civilizing natives" who, missionaries believed, needed reform and a transformation of themselves out of their heathen status. In this overall process, industrial schools became innovative institutions that aimed to integrate native labor into the modern capitalist world by skilling them in industrial trades and to blend the industrial spirit of missionaries with educational and social concerns.

In India, the Western Christian mission societies flourished and expanded under the gaze of British imperialism with the 1813 renewed Charter of the East India Company, which opened up the company territory for missionary activities. Although these societies were established with the aim of spreading the principles of Christianity among the colonized population, their role in constituting and sustaining the imperial civilizing mission is undeniable.[5] One of the aims of the civilizing mission—a discourse to legitimize the presence of imperial forces in colonies—was to sync the colonized "backward" native with Western modernity and scientific rationality. Education became a powerful tool to achieve this goal, and was exhibited in the early half of the nineteenth century by the Anglicists' ideological dominance in modernizing Indians, Indian culture, and the education system according to Western ideals.[6] Along with their primary goal of religious conversion and preaching in the colonies, missionary societies pioneered establishing a widespread education infrastructure that reached areas and native populations untouched or neglected by the colonial state.[7] The educational institutions of missionaries, often riddled with conflicts over the question of curriculum, Christian preaching, and occasional conversion, were actively encouraged by the colonial state through a grant-in-aid system, which allowed the state to share the cost of education with, or defer it to, nonstate actors.[8]

The historiography on the relationship between Christian missionary societies, education, and the empire has shifted significantly from analyzing the role of Christian missionaries as agents of cultural imperialism, education, and civilizing missions, to exploring their links in forging a hierarchized education system, sharpening colonial social hierarchies, and shaping the socioeconomic discourse on slavery or free wage labor.[9] I briefly refer to the analytical structures of this historiography below to situate my own arguments in this chapter, and to show how missionary activities had impacts outside of the religious, cultural, and social spheres as well.

Scholarly work on Christian missionaries in South Asia has focused on the role of missionaries in establishing educational institutes,[10] providing education to "lower and untouchable castes,"[11] offering socioeconomic improvements to Dalits (ex-untouchables), advancing an anti-caste ideology,[12] and promoting rural reconstruction during inter-War period.[13] Research on missionary industrial education, however, is still in its infancy.[14] Soni and Parinitha Shetty have explored the linkages between Christian mission societies in India (the Church Missionary Society and the Basel Mission Society, respectively) and industrial education to show how the industrial pedagogical work of mission societies had a multitude of aims: sustaining the finance of the institution through their labor and produce, generating a Christian civilizing discourse, and disciplining the native body in work ethics.[15] Civilizing missions and the expansion of Christianity to colonies became a means of propagating free trade, opening up colonial markets to Western goods, and producing colonies as spaces of raw material production.[16] Andrew Porter argues that early Victorian missionaries (1830s to 1860s) raised the slogan of "commerce and Christianity," but that by the 1870s the links between Christianity and commerce were marginalized and the idea that missions "should sedulously avoid all economic activity" gained ground.[17] However, missionaries in India and Africa continued to entrench their role in the relationship between Christianity, education, and labor. John and Jean Comaroff's detailed studies of missionaries in South Africa between the 1820s and 1920s show that, through Christian community formation, missionaries cultivated the values of thrift, honest labor, sobriety, and moral restraint among the "heathens" and trained them into being docile laborers, producers of raw material, and ideal consumers, leading to the formation of what they call the "Christian Political Economy."[18] Discussing industrial education in the settler colonies of the British Empire, such as Natal, Western Australia, and the West Indies between 1833 and 1880, Rebecca Swartz argues that as ideas of racial difference hardened in this period, the discourse of civilizing Indigenous people came to encompass pedagogical practices of training Indigenous people in useful labor. Swartz persuades us to look at missionary education as a form of "ambivalent humanitarianism." The imperial and Western discourse of industrial education and labor was couched in the language of civilizing Indigenous people, improving their overall moral and physical state. However, at the local level, it was also enmeshed with settlers' need to control the land and labor of Indigenous people.[19]

This chapter adds to this historiography, arguing that the British Empire, Christian missionaries, and colonial capitalism came to be interlinked through the institution of the mission industrial school. Although the source material and arguments of this study are foregrounded in the late colonial Indian period, the SPG's mission station in Cawnpore was representative of a broader global trend in which missionaries from Western imperial nations operated as a global network in various parts of the colonized world. The missionary education model imposed by the SPG and other Protestant missionary societies adopted a transnational character, with ideas about education and labor circulating across and between missionary stations and other mission societies.[20]

The SPG Mission and Native Labor

The SPG had a history of entanglement with slavery. It is important to refer to the SPG's checkered history to show how the mission transformed its position vis-à-vis colonized labor. Long before its India operations, the SPG acquired the slave-run Codrington sugar plantation in Barbados in 1712 through a bequest, and operated it throughout the eighteenth century until it became unprofitable to buy and maintain slaves. Newly bought and seasoned slaves from Africa worked at the SPG's Codrington sugarcane plantations, built houses for the mission, including the Codrington College, and worked as domestic laborers.[21] In her recent work, Katharine Gerbner demonstrates that Protestant Christians in the Atlantic world, including those working for the

SPG, justified the compatibility of slave labor with their Christian faith. SPG missionaries reinforced this view by quelling the concerns of white planters, who feared that conversion of enslaved peoples to Christianity would lead to their emancipation. In response, the SPG stressed that conversion led only to "Christianized slavery" not to social and economic emancipation.[22] The spectrum of inequality along the axis of spirituality and socioeconomic conditions was also extended to colonial India, where the SPG and other Protestant missionary bodies dealt with agrarian laborers.

The SPG shifted its orientation toward India only in the 1820s when it commenced its proselytization work in Madras in 1825, in Bengal in 1829, and in Bombay in 1830.[23] It established its mission in Cawnpore in 1833, coincidentally the same year the British abolished slavery in their empire. After slavery was abolished, we see the rise of a new post-slavery labor regime, the indentured labor system, and the expansion of "free wage labor" ideology. In India, missionaries engaged with the poorest "lowest caste" and Dalit Indians who were mainly agrarian laborers in North and South India. Historically, many Dalits—who either worked as bonded agrarian laborers in the fields of upper-caste landed elites or performed the most stigmatized menial labor, such as leatherworks, sanitation work, and burying of the dead—converted to Christianity because they believed it would offer better life prospects, including access to formal education and better housing.[24] Missionaries, who only found success in converting the lower orders of Indian society, were critical of the caste-based agrarian servitude, but maintained that Dalits and lower castes should continue to do manual labor.[25] By now, the SPG and other Protestant missionary bodies had evolved to become powerful proponents of free wage labor.[26]

On the other hand, Christian missionaries distinguished between "real" spiritual conversions and conversions made strictly for material gains. As missionaries themselves established coffee plantations with the labor of their converts or persuaded converts to become wage laborers on plantations managed by European capitalists, converts realized that they had not gained a significant socioeconomic leap but were merely shifted from feudal agrarian slavery managed by local landed elites to another system of wage labor bondage controlled by European capitalist planters.[27] Rupa Viswanath, in her work on the relationship between Protestant missionaries (the SPG, the Wesleyan Methodists, the American Arcot Missionaries) and Dalits in the Madras presidency (then called Pariahs), argues that missionaries were only interested in "replacing the *harsh and cruel* oriental slavery with *gentler* free servitude."

They did not make any fundamental challenge to the socioeconomic basis of India's social hierarchy.[28] What this historiography suggests is that Protestant missionaries, though critical of the caste system and agrarian servitude in India, were largely comfortable with not disturbing the caste-based division of mental and manual labor. The lowest castes essentially remained laborers, though they could now labor in a culturally "dignified" sense, as labor itself was proposed as a noble Christian activity.

It was through industrial missions in the late nineteenth and twentieth centuries that Christian missionaries crystallized their relationship with the native "waged" labor. Through these missions, local converts were engaged in industries, crafts, or agricultural activities that directly supported the local Christian mission station, and, in the process, developed a global missionary industrial world where commerce, trade, industrial production, employment generation, and skill development went hand in hand. Pioneered by the German Basel Missionaries in the mid-nineteenth century, industrial missions were looked upon with suspicion by other missionaries, opening up a debate among missionaries over whether the objective of missionaries was to preach the gospel or to train carpenters, smiths, and farmers.[29] However, this debate was settled by the last decade of the nineteenth century with the majority of Christian mission societies having established industrial missions all over the world. Various missionaries articulated their views on industrial missions in missionary journals, conferences, and reports, highlighting the trans-colonial missionary engagement on the topic.

The SPG Expansion and the Birth of the Cawnpore Mission Industrial School

The Cawnpore city, lying between Delhi and Allahabad, was a small outpost of the SPG, but an emerging manufacturing and commercial center. The SPG and the American Methodist Missionaries were some of the mission societies that worked in this region. Cawnpore, being a business and cantonment center, had many Europeans (about 965 in 1857) who saw the SPG providing institutionalized access to Christianity in the city. However, Christian missionary societies, including the SPG, found it hard to covert upper-caste Hindus and Muslims. To gain influence among them and initiate discussions, missionaries used the strategies of establishing literary schools and bazaar preaching.[30] Missionaries believed that educational institutions, the truthful nature of Christianity, and missionary preaching would attract "respectable" natives from socially elite and influential castes. In the 1830s, the SPG operated five general literary schools (attended by 170 boys and 22 native Christians in 1833) that taught subjects such as languages, history, and arithmetic.[31] This strategy of establishing schools for the children of elite-caste natives did not bear fruit, as only a few upper-caste members converted to Christianity.[32] It was famine times, when Christian missionaries organized relief programs for the poor, that brought the lowest and the oppressed social castes closer to missionaries. After the famine in the Bundelkhand region in 1835, the SPG established its first female orphanage, Savada Kothi, in the cantonments.[33]

However, the 1857 anticolonial revolt of Indians against the East India Company rule, which also targeted Christian missionaries, disrupted the activities of the SPG. Historian Rudrangshu Mukherjee shows that Cawnpore was one of the prominent centers of Indian rebellion and violence that killed a number of British personnel, including missionaries.[34] It was only by 1890 that the SPG was able to recreate its local Christian community, with its activities centered in a mission compound located in the city. The SPG ran six vernacular schools across the city with 240 students, one Middle Anglo-Vernacular school in Generalganj with 200 students, and a degree college named the Christ Church College, which offered BA, MA, and LLB courses. The higher educational institutions catered to the whole population of the city and were mainly attended by the affluent Hindu and Muslim families, with a small number of Christian converts reaching the upper echelons of the SPG education system. In 1890, the Christ Church College had 222 Hindus, 45 Muslims, and just 8 Christian students.[35] Orphans and children of the poor converted Christians attended primary schools, orphanages, and sometimes the Anglo-vernacular Middle school.

The SPG's Cawnpore Mission Industrial School owed its origins to an Orphanage Industrial Home, set up between 1889 and 1890, which housed and trained eleven orphans and poor children (usually the sons of converts). This Industrial Home was renamed St. Martin's Industrial School in 1906.[36] The 1897 famine caused a significant growth in the number of students, many of whom were recently orphaned.[37] A series of monsoon failures and the subsequent drought had affected the whole of North India. Crops had failed and rivers had dried.[38] Famine suffering extended to the whole North-Western Provinces of Agra and Oudh, Bombay, Bengal, and the Central Provinces. The government soon opened poorhouses and relief works and employed starving people as laborers on railway lines.[39] Famines caused a significant number of children to be abandoned by their parents, who traveled in search of food and perished on the way. The state authorities placed the abandoned children with the nearest kin or in private orphanages that received support from the government.[40]

Christian missionaries traveled to the worst-hit areas, collected abandoned children and adults, and gave them food and shelter. In a retrospective study of the famine, it was calculated that no fewer than seven thousand children were abandoned, eight hundred of whom were received by the Roman Catholics, American missionaries, and the Church Missionary Society.[41] The SPG missionaries in Cawnpore collected about sixty-six boys and fifty-six girls from the worst hit districts of Allahabad, Delhi, Fatehgarh, Faizabad, Lucknow, Jubblepore, and Banda. Many children died on the journey to Cawnpore. The survivors joined the existing fifty-five pupils already resident in the Industrial Home. In the absence of welfare laws, the colonial government paid the missionaries a maintenance fee of two rupees to per orphan.

Large numbers of orphans falling in the hands of Christian missionaries created alarm among Hindu and Muslim religious advocates. The vernacular newspapers of the time were full of rumors that Christian missionaries kidnapped Hindu and Muslim children. The *Rahbar* newspaper, published in Moradabad, alleged that a train arriving at the Ajmer Railway Station on March 5, 1897, contained a coach full of famine-stricken Hindu and Muslim boys from the north.[42] *Alam-i-Taswir* (published in Cawnpore) reported that missionaries had succeeded in turning 113,000 Hindu and Muslim orphans into Christians by 1885.[43] The *Rohilkhand Gazette* (published in Bareilly) published a highly sensational article in its April 16, 1897, edition, claiming that famine-stricken parents sold their children to missionaries for two annas (the local currency).[44] Such sensational news stories highlighted three themes intermittently: first, they accused missionaries of abducting Hindu and Muslim children and keeping them by force; second, they appealed to Hindu and Muslim elites to build their own orphanages; and, finally, they accused the government of favoring Christian missions by allowing them to acquire famine orphans. In her essay on orphaned childhood in colonial India, Soni captured these anxieties and struggles between Christian, Muslim, and Hindu organizations to take custody of orphaned children and to seek financial support from the state for their upkeep.[45]

Before establishing the Cawnpore Industrial Home in North India, the SPG had already been running industrial homes for orphans in South India. SPG's Nazareth Industrial School and the Madras Industrial School in Tamil

Nadu came into existence in the late 1870s as orphanages. The SPG also established other industrial schools in Calcutta, Ahmednagar, and Delhi to teach orphans and the children of poor Christian converts. Likewise, many other missionary bodies also established orphanages and industrial schools throughout India during or after famines. For instance, during the Great Famine of 1876–1877, the Wesleyan Mission established industrial orphanages at Karur, Hasan, and Kolar. Orphanages emerged as post-histories of famines. Some of these orphanages were later turned into industrial schools, while others continued to operate as orphanages. In the United Provinces of Agra and Oudh, the Cawnpore Mission Industrial School became one of the largest private industrial schools, and the closest government industrial school was situated in Lucknow and was established in 1892. By 1898, there were 240 boys in the Cawnpore School, and about fifty more orphans were expected to join from the Banda missionary poorhouse.[46]

Producing Wage Laborers: The Life Inside and Outside the Cawnpore Mission Industrial School

The Cawnpore Mission Industrial School was located inside the SPG mission station compound, and was composed of three large rooms, a cook house, an office for the house father, and a small playground.[47] The compound was the essence of the local Christian community. Among the white male and female missionaries were Indian Christians and their wives and children—who worked in the mission press and other workshops, and attended mission schools, orphanages, and the industrial school—and Indian servants who worked for the missionaries. Describing a similar Christian community life and the setup of the African mission stations, John and Jean Comaroff argue that "the nineteenth-century mission was at once the visible frontage of empire and a cultural borderland in which Africa and Europe face up to each other."[48] Within the SPG compound were also mission workshops dedicated to carpentry, metalwork, and the printing press. SPG missionaries aimed to create symbiosis between religious, community, and industrial life.

Missionary sources mention the everyday life of the industrial school and of its students and staff in a very fragmentary manner, making it difficult to reconstruct the quotidian functioning of the school. We find a missionary newspaper/journal called the *Cawnpore Quarterly Paper*, which reflected on the industrial activities of the mission, along with the SPG's annual reports and a booklet on the St. Martin's Industrial School. Other than this, there are frequent mentions of the school in the annual educational records of the state.

The SPG missionaries designed a meticulous curriculum and divided schooling into three stages. In the first stage, boys attended the school and acquired the skills of reading and writing a local vernacular, the Hindustani language. The literary focus was limited to the extent that students picked up basic skills to read Christian religious literature and to write accounts and letters. In the second stage, they did an apprenticeship in the mission industrial workshops, and, in the final stage, they were either employed in the mission industries or became journeymen (trainees, somewhere between apprentices and full workmen). All three stages were part of industrial school life, and each stage prepared the student for the next. Boys remained in school until the age of thirteen, after which they joined workshops as apprentices.

Reverend F. Westcott, popularly known as "Chhote Sahib," directed the school in the first decade of the twentieth century. Due to his expertise in industrial education, the colonial government asked him in 1902 to serve as a member of the Industrial Education Commission, which enquired into the state of industrial education in India. At a time when colonial officials were debating the nature of industrial schools—whether they should be commercial or educational—the local colonial government, while allocating government grants-in-aid for "proper" industrial schools, praised the Cawnpore Mission Industrial School boys for manufacturing saleable items.[49] The presence of Westcott in a government enquiry committee showed how the SPG missionaries had pioneered the field of colonial industrial education.

There were fundamental differences among missionaries over the mode of industrial training. Some missionaries believed that industrial training should be given to boys in their early childhood, while others believed that smaller children wasted raw materials and hence their industrial training should begin later.[50] Another point of contention was the degree to which the practice of trades needed to be taught in the industrial school curriculum. Westcott became a key figure in missionary circles deliberating on the question of industrial education and its scope, and he was at the center of settling these controversies. At the All-Missionary Decennial Conference in Madras in 1902, an Industrial Work Committee was appointed under the headship of Westcott. Here, several resolutions were taken, including that industrial training of children should be preceded by limited literary education, and that the best method of industrial training was apprenticing students to workshops where they learned the trade under market conditions.[51]

The resolutions of the 1902 All-Missionary Decennial Conference had profound effects on the Cawnpore Mission Industrial School. In 1903, the school syllabus, including the literary section, was revised to make it more practical and tuned to the needs of mission workshops. Initially, only two hours of literary classes (reading, writing, and arithmetic) were taught by Mr. Westcott, but later lessons in geography, natural resources (wood, metal, soil) found in India, and different types of drawing were also introduced. To discourage boys from despising manual labor and seeking clerical employment, studying the English language was discontinued.[52] The reading lessons focused on teaching the value of hard work, honesty, thrift, discipline, sincerity, and loyalty—values which affirmed the superiority of missionary endeavors, Christianity, and Western civilization. During night classes, senior boys narrated moral stories from the Bible to junior boys.

The industrial school controlled what boys were taught, what boys did during the day and night, and how boys worked in industrial workshops. The total control over a child's body in the industrial school was reflective of the emergence of several similar modern institutions that shaped the subjectivity of the "ideal" poor people in this period. Like prisons, regiments, and reformatory schools, mission industrial schools were a disciplinary institution of Michel Foucault's description and, as Christine Whyte remarks in her chapter in this volume, industrial schools often slipped from being philanthropic to carceral.[53] From the school prospectus, we find the following time schedule.[54]

Table 5.1: Daily Schedule of the Students at the Cawnpore Mission Industrial School

5.15 AM	1st bell for getting up
5.50 AM	parade and roll call
6 AM	short service in church
6.15 to 7 AM	physical drill
7.30 AM	preparation for school
9 AM	breakfast
10 AM to 3 PM	classroom lessons in school and manual training
6.30 PM	dinner
7 PM	school again
8 PM	prayers followed by a bell for bed-time rest

Source: *A Short Account of St. Martin's Craft School* (Cawnpore, SPG Mission Press, n.d.), 8–9, USPG Archive, Bodleian Library.

Clocks, bells, gongs, and whistles regulated the temporal rhythm of daily life at the school, preparing the schoolboys for workshop discipline. While teachers watched the movements of the boys in classrooms, a retired Sikh soldier, who had considerable experience of a regimented life, disciplined the boys outside school hours. From the very beginning, the boys were required to eat, sleep, play, and worship on time.[55] In producing discipline, missionaries imposed Western clock-based notions of time and discipline on students and emptied them of their peasant and local times. In the context of eighteenth-century England, E. P. Thompson has shown that Christian schooling, literature, and missionary preachings were very much at the center of producing industrial time discipline during the British industrial revolution.[56] British imperialism transported Western notions of time, days, and calendar to colonies and made time a tool of cultural imperialism. Colonialism, as a range of studies have shown, was not just a territorial project but also a temporal project which aimed to inculcate a sense of order, punctuality, discipline, and time among the "timeless colonized others" who needed to be taught the value of time and streamlined for the labor demands of colonial capitalism, including of missionary enterprises and colonial settlers.[57]

The Industrial School for boys reinforced existing notions of gendered work. While boys were given training in various industrial trades, such as metalworks, carpentry, and printing, girls in the female orphanage industrial school were taught home-based trades such as needlework, stitching, and shoemaking. The SPG missionaries did not report on the details of the female orphanage industrial school as much as they reported on the male industrial school. In the female orphanage industrial school in 1902, there were twenty students learning knitting and shoemaking.[58] Divya Kannan's work on female boarding schools in Travancore shows that Protestant missionaries saw themselves as rescuers of the poor orphaned girls, who were often trained to become Christian wives, separated from the existing Indigenous domestic environment.[59] The SPG attempted to utilize the labor of the orphanage girls and train them by getting them outside factory work (in the form of piece-good work), but still within the physical boundaries of the orphanage.

Workshops—the most important components of industrial schooling—were exclusively male spaces. After graduating from the industrial school, boys were apprenticed in workshops. The first workshop that came into existence in the mission compound was the printing press in 1892. It was attached to the Christ Church Mission Press Limited, a company that underwent voluntary liquidation in 1895 when its assets were acquired by the SPG

Figure 5.1: Young Men Working in a Carpenter's Shop, Cawnpore, India, circa 1914. Source: Wikimedia Commons.

Mission Press. The workshop trained compositors, mechanics, and bookbinders for the SPG's printing press. W. E. Burrows, an English missionary-cum-mechanist, looked after the trade and training. Hiring British master artisans and mechanics for training purposes was the SPG's favored policy, but it was not always financially possible. Carpentry became the second workshop, and was established in 1894 under the supervision of Babu Atma Ram, a Sikh foreman, who was a trained carpenter from the Lahore School of Art. The workshop did carpentry work needed for the missionary compound, the church, and other missionary institutions. Whenever possible, orders from outside were taken.[60]

The school had emerged as a key site of skilling workers for the missionary workshops. With an increased number of students in the school, new workshops in leather and metal were established. In 1903, the school opened a brass foundry with a finishing workshop under the European supervisor W. T. Huett. It produced goods for ecclesiastical purposes, used in the church, and for business. For example, the molders and fitters in the workshop produced metal castings for Cawnpore cotton mills and factories. In 1906, boys made brass lecterns and crosses for churches in England and Madras. Christian missions throughout North India required a large amount of brass work for their churches, and there was no other place where "Christian labor" produced brass work.[61] To inaugurate the foundry workshop on March 3, 1903, the Lieutenant Governor Sir James Digges LaTouche, along with his wife, traveled to Cawnpore from Lucknow on a special train. The Bishop of Lucknow and the Collector (the administrative head) of Cawnpore also attended the function. In his speech, the governor brought up a rather controversial issue. He warned, "in a workshop of this kind there are many matters which require regulation [such as] the time at which boys cease to be apprentices and earn wages for themselves . . . There is a danger that the boys should be overworked and underfed in the interests of private profit."[62] While first highlighting the dangers of overworking children, the governor shifted the conversation and pleased missionaries by saying, "in the hands of the missionaries here there is no such danger, and the experiment now inaugurated will be of great value both to the Indian Christian community and to Government, which has, as you know an Industrial School of its own at Lucknow."[63]

Nevertheless, workshops were run on strict business lines, and the aim was to train converts or prospective converts, who were mainly "low castes" and Dalits, in real market conditions. The apprenticeship period ranged from three years in the press department to five years in the carpentry department and in the foundry and finishing shop.[64] Orders were received from outside, and an annual sale bazaar at the Station Theatre was also organized every December for the general public to buy the workshop goods. Printers and carpenters generally worked from 7:30 A.M. to 5 P.M. daily except for Saturday, which was a half-working day.[65] When there was an abundance of orders, apprentices along with workmen worked longer hours. Any overtime was represented as voluntary. For example, in 1897 the school authorities reported, "of the famine orphans, 16 are now employed in the Mission workshops, 6 as carpenters, 5 in the press, and 5 as leatherworkers. For the most part they display great zeal, some having gone so far as to ask for longer hours of work."[66]

Workshops kept an account of their sales and expenditures. Doing so was in the spirit of running the workshops on commercial lines and maintaining records that were available for investigation by missionary authorities. We have the sale figures of the workshops for the year 1903, which show that the printing press earned Rs. 13,574 and carpentry earned Rs. 12,453.[67] In the absence of expenditure figures, it is difficult to calculate the extent of profits made. These figures do, however, suggest that the workshops had a substantial business footing in the market.

Table 5.2: Number of Apprentices, Journeymen, and Workers, Cawnpore Mission Industrial School

Trades	*Present apprentices no.*	*Working as journeymen*	*Passed out and working at trade*	*Withdrawn from the school*	*Died*	*Total*
Carpenters	17	3	13	13	1	47
Compositors	6	1	9	1	2	19
Printers	7	1	6	1	1	16
Bookbinders	3	0	2	0	1	6
Leather workers	5	0	12	4	2	23
Moulders, Fitters	15	0	4	0	0	19
Total	53	5	46	19	7	130

Source: *A Short Account of St. Martin's Craft School* (Cawnpore, SPG Mission Press, n.d.), 17, USPG Archive, Bodleian Library.

In terms of the scale of mission industries and the production of Christian industrial labor, the SPG in Cawnpore was a relatively smaller player compared to the entrepreneurial activities of the Basel Mission, whose factories employed 3,633 workers, including 2,679 converted workers, in 1913.[68] The table above lists the number of apprentices, journeymen, and independent workers the Cawnpore school produced in 1905.

Since the late nineteenth century, Cawnpore had been growing into a merchant and industrial city with towering cotton mills, leather factories, and mechanical establishments. David Thomas argues that British capital investment along with government support was critical in the commercial rise of Cawnpore.[69] Many of the European traders and company owners were sympathetic to missionaries.[70] Such an urban environment allowed missionaries to forge a close link with employers, who would not only support the church financially but also employ missionary-trained Christian converts. We come to know about the numbers of the Christian converts working in various factories from a report by Westcott. In 1899, Westcott asked managers of factories in the North-Western Provinces and Oudh, Punjab, and Bengal to answer a set of questions. Out of 123 employers of labor, only 49 employers replied. Among those who responded, 23 employed converted Christian labor: a total of 293 workers, 39 as clerks and 254 as workmen.[71] Workmen were divided into the following occupations: 86 as compositors and readers,

73 as fitters and mechanics, 28 as leatherworkers, 7 as bookbinders, and 58 in various other trades. We do not know how many of these workmen were trained at the Cawnpore Mission Industrial School.

The SPG also secured work for the orphanage girls through close ties with local factories. Senior girls of the orphanage worked for the two large factories of Cawnpore—the N.W. Tannery and the Cawnpore Woollen Mills. These two factories had established their workshops in the mission compound—one for the leatherwork and the other for the mechanical knitting. There were twenty girls working in these workshops in 1902. A salaried male mistri (master artisan) trained girls and supervised their work. The leatherwork consisted of pasting together precut pieces of leather or using sewing machines to stitch canvas shoes. The finished work was returned to the tannery for final inspection and sale. The Woollen Mills, established in 1876, produced flannels, blankets, hosiery, scarves, socks, and stockings, and also supplied lacework, machines, and material to the mission. In 1902, the compound contained four machines for making stockings. Factories paid the girls' wages directly to the orphanage superintendent, but only after deducting the cost of material such as cotton, needles, and pastes. The superintendent, after further deducting various incidental expenses, paid wages to the girls according to their work. Missionaries, however, maintained that girls were inefficient and could hardly earn a living, even after working the whole day. As per missionary calculations, to support herself a girl was required to produce ten pairs of stitched shoes, but they could only make six pairs in a day.[72] We can see two interrelated processes working here. First, missionaries were closely involved in the process of feminizing industrial work—that is, certain aspects of industrial work could be done by women based at home, and, in this case, it was the job of stitching the canvas shoes. Second, the domesticizing of female factory work—that is, women could do factory work without physically traveling to factories—ensured that female workers were not recognized as official industrial workers and so could be paid less.[73] This suited the missionary ideal of producing productive female bodies but within the "respectable" boundaries of home.

Conclusion

Through this study, I have shown that industrial missions and industrial education emerged as key global discourses toward the late nineteenth century, allowing colonial Christian missionaries, the SPG in our case, to engage with and champion "free" wage labor of the colonized people. These discourses also positioned the SPG as a modernizing force in the twentieth century, distancing them from their slave owning/slavery-supporting image in the eighteenth century. These industrial missions and pedagogic activities allowed religious missionaries in colonies across Asia and Africa to indulge in industrial pursuits and local industries and to establish mission industrial schools, which posed as supporting poor converts materially and equipping them with industrial skills. However, in the process of training converts and the poor in several trades, the missionary schooling system came to appropriate and push the socially differentiated education structure: practical industrial education for the poor, lowest castes, and orphans, versus liberal education for fee-paying social elites. As outlined in the introduction and in the chapter by Janne Lahti in this volume, such a differentiated structure of education along the lines of race and class was also prevalent in the Atlantic world. Through missionary journals, educational conferences, and global missionary meetings, Western Christian missionaries circulated ideas of industrial and practical education in the empire and outside of it. While the global context of missionary industrial education is stressed in the scholarship, this chapter has highlighted how this globalized education was reliant on not just hierarchies of class and race in localized settings, but also of caste in India.

Industrial schools emerged as innovative institutions that helped to skill socioeconomically and culturally marginalized people, but also teach them lessons in obedience, time discipline, and industriousness that fit well into the logic of industrial capitalism, social difference, and the imperial civilizing mission. I show that SPG and other missionaries tied the global discourse of industrial education intricately with Christian values of hard labor, the notion of a calling to a higher purpose, and civilizing the "native." In the colony, this pedagogic project of reform and transformation was dependent on an idealized picture of undisciplined and irrational colonized bodies who despised having to conduct manual labor for a living. The process of industrial schooling rationalized these notions. Missionary education was seen as a necessity to purge the irrational and uneconomical behavior of natives. As various chapters in this volume show, such notions about improving and rationalizing the socially marginalized poor through industrial, practical, and vocational education were not limited to missionary education. They extended to secular industrial education programs run by the state institutions.

I also argue that industrial education allowed missionaries to build a cohesive Christian community that traversed the religious and material needs of converts. This setup benefited a range of actors, including the missionaries who wanted to run financially self-sufficient mission stations; the converts who sought help from missionaries in times of scarcity and immense poverty; and European industrialists and employers who needed trained, skilled, and well-behaved labor. However, missionary industrial education was cautious to not disturb the local status quo of the fundamental division of mental and manual labor—a strict manual laboring life for the poor, "lowest castes" by training them in industrial schools and an open career-oriented life for local social elites by giving them high literary education. In doing so, missionaries also constituted the field of modern industrial education in India and reaffirmed the essentialized image of "lower castes" as laborers, albeit skilling them in commercially relevant trades.

Notes

I am grateful to Rupa Viswanath for her valuable feedback on an earlier draft of this paper, provided during the course of my doctoral work. I also thank review editors for their valuable suggestions.

1. Gertrude Himmelfarb, *The Idea of Poverty: England in the Early Industrial Age* (New York: Alfred A. Knopf, 1985); Michael Dennis, "Schooling Along the Color Line: Progressives and the Education of Blacks in the New South," *Journal of Negro Education* 67, no. 2 (1998): 142–156; Arun Kumar, "Learning to Dream: Education, Aspiration, and Working Lives in Colonial India (1880s–1940s)" (PhD diss., University of Göttingen, 2017).

2. Max Weber, *The Protestant Ethic and the Spirit of Capitalism* (New York: Routledge, 2001, original 1930); E. P. Thompson, "Time, Work-Discipline, and Industrial Capitalism," *Past & Present* 38 (1967): 56–97.

3. Kirsten Kamphuis and Elise van Nederveen Meerkerk, "Education, Labour, and Discipline: New Perspectives on Imperial Practices and Indigenous Children in Colonial Asia," *International Review of Social History* 65, no. 1 (2020): 1–14.

4. Arun Kumar, "Skilling and Its Histories: Labour Market, Technical Knowledge and the Making of Skilled Workers in Colonial India (1880–1910)," *Journal of South Asian Development* 13, no. 3 (2018): 249–271; Bidisha Dhar, "Technical Education Discourse in India: State and the Artisans, 1880s–1914," *Studies in History* 33, no. 2 (2017): 213–233.

5. The civilizing mission was not a homogenous project; its construction, the mapping of difference between the colonizer and the colonized, and the discourse makers of difference kept shifting throughout British colonialism; Catherine Hall, *Civilising Subjects: Metropole and Colony in the English Imagination, 1830–1867* (Oxford: Polity, 2002), 1–20.

6. Jana Tschurenev, *Empire, Civil Society, and the Beginnings of Colonial Education in India* (Cambridge: Cambridge University Press, 2019), 13–14; Lynn Zastoupil and Martin Moir, eds., *The Great Indian Education Debate: Documents Relating to the Orientalist-Anglicist Controversy, 1781–1843* (Richmond: Curzon, 1999).

7. Tschurenev, *Empire, Civil Society, and the Beginnings of Colonial Education in India*; Larry Prochner, Helen May, and Baljit Kaur, "'The Blessings of Civilisation': Nineteenth-Century Missionary Infant Schools for Young Native Children in Three Colonial Settings—India, Canada and New Zealand 1820s–1840s," *Paedagogica Historica* 45, no. 1–2 (2009): 83–102.

8. Hayden J. A. Bellenoit, "Missionary Education, Religion and Knowledge in India, c. 1880–1915," *Modern Asian Studies* 41, no. 2 (2007): 369–394.

9. For a foremost study on this theme, see Rupa Viswanath, *The Pariah Problem: Caste, Religion, and the Social in Modern India* (New York: Columbia University Press, 2014).

10. Hayden J. A. Bellenoit, *Missionary Education and Empire in Late Colonial India, 1860–1920* (London and New York: Routledge, 2015).

11. Philip Constable, "Sitting on the School Verandah: The Ideology and Practice of 'Untouchable' Educational Protest in Late Nineteenth-Century Western India," *Indian Economic and Social History Review* 37, no. 4 (2000): 383.

12. John C. B Webster, *The Dalit Christians: A History* (Delhi: ISPCK, 2009); G. A. Oddie, "Christian Conversion in the Telugu Country, 1860–1900: A Case Study of One Protestant Movement in the Godavery-Krishna Delta," *Indian Economic & Social History Review* 12, no. 1 (1975): 61–79; P. Sanal Mohan, *Modernity of Slavery: Struggles Against Caste Inequality in Colonial Kerala* (Oxford: Oxford University Press, 2015); Dick Kooiman, *Conversion and Social Equality in India: The London Missionary Society in South Travancore in the 19th Century* (New Delhi: Manohar Publishers & Distributors, 1989).

13. Caste is the ancient, birth-based socioeconomic and religious gradation of society; Harald Fischer-Tiné, "The Ymca and Low-Modernist Rural Development in South Asia, c. 1922–1957," *Past & Present* 240, no. 1 (2018): 193–234.

14. Parinitha Shetty, "Missionary Pedagogy and Christianisation of the Heathens: The Educational Institutions Introduced by the Basel Mission in Mangalore," *Indian Economic & Social History Review* 45, no. 4 (2008): 509–551; Jaiprakash Raghaviah, *Basel Mission Industries in Malabar and South Canara, (1834–1914): A Study of Its Social and Economic Impact* (New Delhi: Gian Publishing House, 1990); Fedrick Sunil Kumar N. I., "The Basel Mission and Social Change—Malabar and South Canara a Case Study (1830–1956)" (PhD diss., University of Calicut, 2006).

15. Soni, "Learning to Labour: 'Native' Orphans in Colonial India, 1840s–1920s," *International Review of Social History* 65, no. 1 (2020): 15–42; Shetty, "Missionary Pedagogy and Christianisation of the Heathens."

16. Andrew Porter, *Religion Versus Empire? British Protestant Missionaries and Overseas Expansion, 1700–1914* (Manchester: Manchester University Press, 2005); Andrew Porter, "'Commerce and Christianity': The Rise and Fall of a Nineteenth-Century Missionary Slogan," *Historical Journal* 28, no. 3 (1985): 597–621; Brian Stanley, "'Commerce and Christianity': Providence Theory, the Missionary Movement, and the Imperialism of Free Trade, 1842–1860," *Historical Journal* 26, no. 1 (1983): 71–94; Eric Thomas Stokes, *The English Utilitarians and India* (New York: Oxford University Press, 1989).

17. Porter, "'Commerce and Christianity,'" 617.

18. John L. Comaroff and Jean Comaroff, *Of Revelation and Revolution: The Dialectics of Modernity on a South African Frontier*, vol. 2 (Chicago: University of Chicago Press, 1997), 9–15, 2195–2207.

19. Rebecca Swartz, *Education and Empire: Children, Race, and Humanitarianism in the British Settler Colonies, 1833–1880* (Switzerland: Palgrave Macmillan, 2019).

20. Kim Christiaens, Idesbald Goddeeris, and Pieter Verstraete, eds., *Missionary Education: Historical Approaches and Global Perspectives* (Leuven: Leuven University Press, 2021).

21. J. Harry Bennett, Jr., "The Problem of Slave Labor Supply at the Codrington Plantations," *Journal of Negro History* 37, no. 2 (1952): 115–141; Stephen Bates, "Church Apologises for Benefiting from Slave Trade," *Guardian*, February 9, 2006, www.theguardian.com/uk/2006/feb/09/religion.world.

22. Katharine Gerbner, *Christian Slavery: Conversion and Race in the Protestant Atlantic World* (Philadelphia: University of Pennsylvania Press, 2018), 123–48; 66–67 (Ebook).

23. C. F. Pascoe, *Two Hundred Years of the S.P.G.: An Historical Account of the Society for the Propagation of the Gospel in Foreign Parts, 1701–1900 (Based on a Digest of the Society's Records.)* (London: Published at the Society's Office, 1901).

24. Viswanath, *The Pariah Problem*; Webster, *The Dalit Christians.*

25. Viswanath, *The Pariah Problem.*

26. Dick Kooiman, "Conversion from Slavery to Plantation Labour: Christian Mission in South India (19th Century)," *Social Scientist* 19, no. 8/9 (1991): 57–71.

27. Kooiman, "Conversion from Slavery to Plantation Labour."

28. Viswanath, *The Pariah Problem*, 88–89, 15–19, 40–99.

29. R. Wardlaw Thompson, "The Need for Industrial Mission," *The East and the West: A Quarterly Review for the Study of Missions*, vol. 2 (Society for the Propagation of the Gospel in Foreign Parts, 1904), 402; on the industrial activities of the Basel missionaries, see Raghaviah, *Basel Mission Industries.*

30. For a detailed analysis of these methods, see Arun Kumar, "The 'Untouchable School': American Missionaries, Hindu Social Reformers and the Educational Dreams of Labouring Dalits in Colonial North India," *South Asia: Journal of South Asian Studies* 42, no. 5 (2019): 823–844.

31. Herbert Moore, *Here and There with the S.P.G. in India* (Westminster: Society for the Propagation of the Gospel in Foreign Parts, 1905), 41.

32. On education-led conversions and local uproars surrounding such conversions, see Bellenoit, "Missionary Education, Religion and Knowledge in India."

33. H. R. Nevill, *Cawnpore: A Gazetteer of the District Gazetteers of the United Provinces of Agra and Oudh* (Allahabad: Superintendent Government Press, 1909), 122.

34. Rudrangshu Mukherjee, "'Satan Let Loose upon Earth': The Kanpur Massacres in India in the Revolt of 1857," *Past & Present*, no. 128 (1990): 92–116.

35. *The Mission Field 1890* (London: G. Bell and Sons, 1890), 139–140, United Theological Archives (UTC).

36. Pascoe, *Two Hundred Years of the S.P.G.*, 599c; SPGFP Annual Report, 1903, 91.

37. J. B. Lyall et al., *Report of the Indian Famine Commission 1898* (Shimla: Government Central Printing Office, 1898), 314.

38. For data on rain and crop failure in these years, see: *Resolution on the Administration of Famine Relief in the North-Western Provinces and Oudh During 1896 and 1897* (Allahabad: North-Western Provinces and Oud Government Press, 1897), 14–17.

39. Sanjay Sharma, *Famine, Philanthropy and the Colonial State: North India in the Early Nineteenth Century* (New Delhi: Oxford University Press, 2001); Madhavi Jha, "'Men Diggers and Women Carriers': Gendered Work on Famine Public Works in Colonial North India," *International Review of Social History* 65, no. 1 (2020): 71–98.

40. Lyall, *Report of the Indian Famine Commission*, 316; Soni, "Learning to Labour," 24–25.

41. Rev. J. R. Hill, "The Indian Famine—A Retrospect," in *The Mission Field 1899* (London: G. Bell and Sons, 1899), 88, UTC Archive.

42. *Selection from the Vernacular Newspapers Report (SVNR) North Western and Oudh* (March, 1897), 228, IOR/L/R/5/74, British Library.

43. *SVNR*, November, 1897, 704, IOR/L/R/5/74, British Library.

44. *SVNR*, April, 1897, 283, IOR/L/R/5/74, British Library.

45. Soni, "Learning to Labour," 25–26.

46. Westcott, "Famine in India–Relief in Cawnpore."

47. Boys' Orphanage, *Cawnpore Quarterly Paper (QPC)*, October, 1897, 3–4; Moore, *Here and There*, 46.

48. Comaroff and Comaroff, *Of Revelation and Revolution*, 293.

49. File No. 268 A, Rules for Grants-in-aid to Industrial School in the North Western Provinces and Oudh. July, 1900/ IOR/ 5822.

50. "Notes on Industrial Education in India," *Harvest Field*, February, 1904, 47, UTC Archive.

51. *Report of the Fourth Decennial Missionary Conference Held in Madras December 11th–18th, 1902* (London: Christian Literature Society, 1903), 141–143, https://archive.org/details/reportoffourthde00indi.

52. *Annual Report of the Society for the Propagation of the Gospel in Foreign Parts* [hereafter *SPGFP*], 1904, 90.

53. Michel Foucault, trans. Alan Sheridan, *Discipline and Punish: The Birth of the Prison* (London: Vintage, 1995).

54. *A Short Account of the St. Martin's Craft School* (*A Short Account*), 8–9.

55. "St. Martin's Home, Cawnpore," *The Mission Field 1913* (Society for the Propagation of the Gospel in Foreign Parts, 1913), 103.

56. Thompson, "Time, Work-Discipline, and Industrial Capitalism," 56–97.

57. Keletso Atkins, *The Moon Is Dead! Give Us Our Money!: The Cultural Origins of an African Work Ethic, Natal, South Africa, 1843–1900* (Portsmouth: Heinemann, 1993); Giordano Nanni, *The Colonization of Time: Ritual, Routine, and Resistance in the British Empire* (Manchester: Manchester University Press, 2012); Arun Kumar, "3 The Nights of Bombay Workers (1870–1920)" in *Working at Night: The Temporal Organisation of Labour Across Political and Economic Regimes*, ed. Ger Duijzings and Lucie Dušková (Berlin: De Gruyter Oldenbourg, 2022), 45–68.

58. Lieutenant-Colonel J. Clibborn, C. A. Radice, R. E. Enthoven, and Rev. F. Westcott, *Report on Industrial Education: Part II: Proceedings of Conferences* (Calcutta: Superintendent of Government Printing, 1903), 30.

59. Divya Kannan, "Educating Poor Girls: The London Missionary Society in the Nineteenth Century South India," in *Religion and Poverty*, ed. Frank Jacob (Berlin: Neofelis Verlag, 2015), 52–70.

60. *SPGFP*, 1906, 91; *A Short Account*, 12.

61. *SPGFP*, 1906, 91.

62. *CQP*, April, 1903, 13.

63. *CQP*, April, 1903, 13.

64. *Directory of Technical Institutions in India Including Agricultural, Industrial, Technical, Commercial and Science Schools and Colleges* (Allahabad: Standard Press, 1909), 162.

65. *The Mission Field 1913*, 105.

66. *CQP*, April, 1897, 4.

67. *CQP*, April, 1897, 14.

68. *Report of the Basel Mission*, 1913, 20, UTC Archive.

69. David A. Thomas, "Lucknow and Kanpur 1880–1920: Stagnation and Development Under the Raj," *South Asia: Journal of South Asian Studies* 5, no. 2 (1982): 68–80.

70. Pascoe, *Two Hundred Years of the S.P.G.*

71. *CQP*, October, 1900, 1–6.

72. *CQP*, October, 1902, 6–7.

73. For an interesting study of women lace makers and the domesticization of lacemaking, see Maria Mies, *The Lace Makers of Narsapur: Indian Housewives Produce for the World Market* (North Melbourne: Spinifex Press, 2012), especially Chapter 3.

CHAPTER 6

"A Good Prison Is Like a School": Industrial Education in Schools and Prisons in the Philippines

Sarah Steinbock-Pratt

Introduction

In 1914, O. Garfield Jones, a teacher in the General Office of the Bureau of Education, declared in the *Philippine Craftsman* that while a "century ago a school was like a prison," today a "good prison is like a school." Indeed, he continued, Bilibid Prison was a model school, closer in some ways to "the pedagogical ideal than the ordinary school." Although the purpose of schools was "*formatory* and not reformatory," Jones argued that the schools and prisons shared the twin goals of creating good citizens and "real producers."[1] While most educational officials would perhaps not have bragged about the similarity between public schools and prisons, they were deeply interested in these goals, paying close attention to Bilibid's industrial shops and attempting to learn from its commercial success. Especially during the heightened attention to industrial education in the 1910s, Philippine public schools were increasingly depicted as businesses, while prisons were likened to schools. This rhetoric of tutelage and reform, and the reality of industrial training and commercialization, highlights the intersection of colonial pedagogy and penology, and their shared purpose of shaping comportment and securing labor suited to U.S. imperial designs.

There was a clear connection between the work of the Bureau of Education and the Bureau of Prisons. The connections between racial capitalism, industrial education, and prison labor have been well established by other scholars.[2] These linkages extended to the colonial Philippines. Historians have also examined the ways in which Philippine penal institutions were used as laboratories for colonial reform and as institutions for securing and producing a steady supply of labor.[3] Yet the links between prisons and public schools—especially in the realm of industrial education—have not been adequately explored.[4]

An examination of industrial work in both schools and prisons in the colonial Philippines demonstrates that these institutions did not develop in isolation, but rather were intertwined, being formed, justified, and praised with similar impulses and rhetoric. Beyond this, U.S. colonial schools and prisons looked to one another for models of how to train and control workers. From 1905 to 1916, public schools and prisons were even under the same administrative department, part of the Bureau of Public Instruction, with the Secretary of Public Instruction responsible for the reformatory projects in both institutions.

Schools and prisons were, of course, very distinct institutions, with drastically different levels of control. Unlike the context of the HMS *Mars* or the U.S. Indian Schools explored by Christine Whyte and Janne Lahti, respectively, in this volume, teachers had a limited ability to compel attendance and dictate comportment within the classroom, and the curriculum was shaped by input from educational officials, influential Americans, and Filipino politicians in Manila, and from teachers, Filipino elites, and community members in the provinces. Yet U.S. colonial prisons and penal colonies often utilized the same rhetoric of uplift, reform, and tutelage as the public schools and, especially by the 1910s, displayed a similar concern with the productivity of Filipino bodies. In attempting to control, discipline, and direct learning and labor, these institutions studied and borrowed from each other. Looking at these connections reveals the delicate mechanisms of colonial control; the successful use of suasion, often in the form of a promise of economic opportunity, as well as coercion, in the attempts to form pliant and productive colonial workers.

The depiction of Filipinos as children in need of tutelage placed schools, as both metaphor and reality, at the heart of the U.S. colonial project in the Philippines.[5] Advocates for empire used education to justify annexation, arguing that it would prepare Filipinos for eventual independence. Industrial education became an important part of this calculus, depicted by its boosters as the vehicle that would drive economic development, prepare Filipinos for productive labor, and usher in a modern economy and workers. Other apparatuses of state control, including penal institutions, also used this rhetoric.

The tuition provided by schools and prisons, and the preparation for full citizenship they claimed to offer, was held up by the colonial state as proof of the benevolence and efficacy of U.S. empire.

Overseas empire was a contested project from the beginning, in both the United States and the Philippines. While the U.S. had a number of objectives in seizing territory in the Pacific, demonstrating that U.S. empire was based on "benevolent assimilation," a term coined by President William McKinley, was a real and concrete goal. It was the lynchpin of a strategy to try to reconcile Americans wary of an imperial republic and to win the cooperation of Filipino nationalists. Prisons and schools became linked through this objective, serving the twin agendas of tangible colonial benevolence and economic development linked to U.S. industrial capitalism. If these institutions were used to justify empire, however, they were also used by nationalist Filipinos to push for independence. They argued that the much-touted success of colonial reform in schools and prisons was proof that Filipinos had learned the lessons that the U.S. had come to teach and were ready to run their own affairs.

This chapter will first sketch out the beginnings of U.S. empire in the Philippines and the precedents to educational and penal institutions that were created in the islands. It will then examine the different types of industrial training, with a particular focus on the relationship between handicraft and shop work in the public schools and Bilibid. Finally, it will highlight agriculture as a focus of industrial programs, and the similar imperatives and limitations of agricultural programs in schools and at Bilibid and the Iwahig penal colony.

Precedents to Industrial Training

The United States annexed the Philippines in 1899 after its victory over Spain in the Spanish-American War. This sparked the Philippine-American War, in which Philippine nationalists continued the fight for independence. While the war was ongoing, the United States took steps to establish a civilian government, including a public school system with almost a thousand American teachers. Civilian and military officials both hoped that free public education would help pacify the Philippines and win support for the colonial state.[6]

From the beginning, the public school system was a subject of contention, as colonial officials, teachers, students, community members, and politicians debated the purpose and ideal form of education for the Philippines. In the early years, the struggle over education largely focused on whether an academically or industrially centered curriculum would create a model school system. Building on a discourse that derided both Spaniards and Filipinos as fundamentally lazy and despising manual labor, advocates of industrial training argued that Filipinos needed to be taught how to work. This discourse, in turn, was part of a global imperial and racialized narrative depicting colonized peoples as unwilling to work, which was part of an effort to compel and control labor, as Dolf-Alexander Neuhaus's chapter also demonstrates. At its heart, industrial education was meant to rehabilitate "problem" populations by incorporating them into U.S. industrial capitalism. As such, schools for non-Christians, especially those for Igorots and Moros, embraced industrial instruction before schools for Christian Filipinos.[7] By the 1910s, the industrial model, including handicrafts, manufacturing, farming and agriculture, and housekeeping, became central to colonial education, though academic instruction continued throughout this period. American investors, colonial officials, and teachers looked to industrial education as vital for the economic development of the islands and the training of workers. After 1925, the Bureau of Education shifted from an industrial to a vocational model, which stressed production for the home and preparation for careers as skilled craftsmen and farmers. While similar in its essential goals, this approach was embraced by nationalist Filipinos looking to eventual independence and economic self-sufficiency. Indeed, as the chapters by Dolf-Alexander Neuhaus and Bronwen Everill demonstrate, local elites often co-opted the rhetoric and policies of industrial education for their own ends.

Models and pedagogies of industrial education traveled around the globe in the nineteenth and early twentieth centuries. U.S. metropolitan colonial educational policies influenced officials in the Philippines, especially those which aimed to control, reform, and uplift African American and Native American children.[8] Politicians and educational officials explicitly looked to institutions like Hampton, Tuskegee, and Carlisle when envisioning and planning schools for the islands. At base, industrial education aimed at reforming pupils into productive workers integrated into a global capitalist marketplace that served the perceived needs of local economic development and U.S. industrial power.[9] At the same time, conditions in the Philippines were sufficiently distinct as to make education policies and actual schools very different from other colonial schools.[10] Schools in the Philippines were less coercive and violent than the institutions examined by Lahti and Whyte in this volume. This was due to the desire both to depict U.S. colonialism as benevolent and to encourage Filipino cooperation in colonial education. Moreover, by the time industrial education became a primary focus of colonial education in the mid-1910s, the Jones Act had been passed, establishing a Philippine Senate and holding out the promise of eventual independence. This gave Filipino politicians, community leaders, parents, and students more leverage in shaping colonial education than was the case with most colonial schools for nonwhite and poor children elsewhere.

Spanish precedents also influenced the shape of U.S. colonial schools and prisons. Spain began investing more heavily in public schools on the islands in the mid-nineteenth century, though these schools only ever enrolled a small percentage of Filipino youths.[11] The U.S. colonial state also took over penal systems established by the Spanish, including Bilibid Prison and the penal colonies of San Ramon in Zamboanga, and Iwahig on Palawan.[12] In its reincarnation, Iwahig was modeled after an industrial reform school in New York, and its population was made up of prisoners from Bilibid prison in Manila.[13] That a penal colony would take a reform school as its standard highlights the connections between schools of thought on penology, pedagogy, and colonialism. Thomas Mott Osborne, a well-known advocate of prison reform, declared that imperialism itself was "only the reform school principle on a larger scale."[14] While both school and prison officials looked to U.S. institutions to conceptualize the work they would do in the Philippines, once these institutions were established, they began to draw inspiration from models closer to hand, and to be influenced by intracolonial examples of reform and control.

Industrial Education

While the first general superintendent of public instruction was initially convinced that schools in the Philippines should be developed along the lines of Hampton and Tuskegee, he did little to further industrial education in the islands, due to the Filipino preference for classical training and because the early focus was on English language and primary instruction.[15] During the tenure of the second general superintendent, David P. Barrows, the major battles over the ultimate purpose and direction of education were fought. While Barrows favored a primarily liberal arts education, he instituted a limited industrial training program, which he described with similar rhetoric to that used to justify Native American educational policy.[16] Contrasting white American and Filipino children, he argued that instead of counterbalancing materialism, Philippine education needed to provide "a stimulus to 'practical' activity."[17] He added industrial and manual instruction to the curriculum, including gardening and simple tool work for the boys, and basic housekeeping and sewing for the girls, with increased industrial work in the intermediate schools.[18] Barrows also directed division superintendents to keep the "element of salability" in mind when administering industrial programs.[19]

The first serious and sustained attempts at industrial training were initiated in schools for non-Christian children, who were viewed by American officials as lower on the civilizational ladder and therefore in even greater need of manual training. The educational system was bifurcated from the beginning, with separate schools and pedagogical approaches envisioned for Christian and non-Christian students.[20] Within that division, moreover, colonial educational officials often made finer distinctions of the type of education believed appropriate for non-Christians, including Igorots, Negritos, Moros, and Lumad peoples on Mindanao. A debate ran through the early colonial period, for example, about whether barrio or boarding schools were preferable for Igorot children, with Directors of Education oscillating between the two.[21] Generally, however, the schools established for non-Christians focused much more heavily on industrial education.[22] The formation of separate administrative regions for non-Christian areas of the Philippines systematized and accelerated this approach.[23] Schools for non-Christians, then, served as incubators for industrial education writ large.

Industrial education was introduced to Moro schools very early on. By 1902, Emerson Christie, the head of the newly created Moro Industrial School, set his students to producing native hats, mats, turbans, and wood carvings to sell as curios to American tourists and military and civil officers. He reported that the articles produced at the school were so "piquantly Moro and original" that the children could not make them fast enough to keep up with demand.[24] Early forays into commercialized production like this demonstrated to colonial officials that industrial education could be financially lucrative.

Industrial education was prioritized and centralized under Frank R. White, the general superintendent from 1909 to 1913.[25] He issued a revised curriculum in 1909 that included gardening, weaving, basketry, bamboo and rattan work, sewing, and lace and embroidery.[26] White also created a department to oversee industrial education in the schools, and to ensure that handiwork was done with the goal of making "a serviceable and salable article."[27] Every trade and farm school, moreover, would be operated like a business.[28] To ensure uniformity, the Bureau issued an avalanche of bulletins giving precise instructions

Figure 6.1: Basketry class, Samar, 1912 (Record Group 350-P, National Archives and Records Administration, College Park).

and blueprints. From 1912 to 1917, it also published *The Philippine Craftsman*, a magazine devoted to industrial work in the schools.[29] The Bureau believed (erroneously, as it turned out) that items like hats, baskets, embroidery, and lace not only could be sold in the Philippines, but that there would be "an enormous market" for these products in the United States as well.[30]

Handicraft work also had the advantage of visibility. Hats, baskets, embroidery, and lace could all be displayed, handled, taken home, photographed, and printed in official reports and booklets. These products were tangible proof of the progress of Philippine public schools. The Bureau of Education devoted much of its time to securing distinctive, "national, *Filipino*" handicrafts made in the schools for national and international fairs, especially the 1915 Panama-Pacific Exposition.[31] Adapting designs from handicrafts made by non-Christians was also part of an effort to create a signature look for Philippine industrial products to increase the popularity and salability of these articles. The very first page of the exposition booklet published by the Bureau of Education had a photograph of an abaca coiled basket with a "Manobo" lizard design.[32]

The Bureau of Education was particularly enthusiastic about the prospects for Philippine goods because of the ongoing world war.[33] Bolstered by

THIRTEENTH ANNUAL REPORT, BUREAU OF EDUCATION—1913.] PLATE XIV.

A MANDAYAN SKIRT CLOTH FROM MINDANAO.

This cloth contains a wealth of primitive design units including the "woman" figure employed in the illustrations below.

DESIGN FOR COILED WORK BASKET USING THE "WOMAN" UNIT FROM MANDAYAN SKIRT CLOTH.

A working design adapted from the Mandayan skirt cloth.

COMPLETED WORK BASKET OF COILED ABACA ON SEDGE FOUNDATION WITH "WOMAN" DESIGN FROM MANDAYAN CLOTH.

This series of three illustrations shows the method used in adapting Philippine designs for school industrial products. The study of typical Philippine designs for the industrial lines in which the schools are engaged is an important and interesting work. Design units from native articles, many of them of antique origin, are employed in natural and conventionalized forms.

Figure 6.2: Illustration showing the process of creating an "authentic" Philippine basket (Thirteenth Report of the Director of Education, 1913).

this opening, in 1916, a General Sales Department was created within the Bureau of Education.[34] Now the work of establishing industries, purchasing products, and finding markets was the jurisdiction of the Bureau of Education. This meant, as W. W. Marquardt, the Director of Education, declared, that he must assume "the role both of a pedagogue and of a business man."[35]

As schools were becoming businesses by the 1910s, prisons were becoming schools. In 1905, the Bureau of Prisons was transferred from the Department of Commerce and Police to the Bureau of Public Instruction, to balance the duties of each department and to stress the "educational feature" of the prison.[36] At that point, then, the Secretary of Public Instruction was responsible for the reformatory colonial projects of both the public schools and the national prisons. Education was central to the rehabilitative efforts at Bilibid; a prison school was opened in 1909, where inmates learned English, basic arithmetic, literature, and history. Attendance was compulsory for everyone under fifty.[37] The teachers in the school were former public-school teachers who were currently serving sentences.[38]

The Director of Prisons also put compliant, long-term prisoners to work in the industrial division, which quickly became one of the most highlighted features of the prison.[39] By 1907, the industrial work in the prison had been completely revamped, and included divisions in machine work, hardwood, construction, rattan and bamboo furniture, silversmithing, tailoring, laundry, and lace and embroidery (for female prisoners). A new accounting system allowed prison officials to track expenditures and profits for everything being made.[40] By 1908, there were 894 prisoners working in Bilibid's industrial shops.[41] This work could progress much more quickly than in the public schools. There were no recalcitrant students—recalcitrant prisoners could be punished or removed from the industrial division—no concerned parents to deal with, and no need to gain community buy-in. In this sense, Bilibid was the ideal laboratory for the colonial experiment of industrial work and reform.[42] And the Bureau of Prisons trumpeted the work of industrial division, declaring in 1907 that its work was important "both from an economic and a reformatory standpoint."[43]

The commercialization of industrial work at Bilibid predated that of the public schools, with the exception of the Moro schools. As early as 1905, a prison salesroom was completed, where local and international visitors could view and purchase items made by inmates.[44] And visitors flocked to Bilibid, not just to buy furniture and other goods, but to tour the prison itself, see the prisoners do military-style drills, and enjoy the spectacle of American

Figure 6.3: Rattan furniture shop, Bilibid, circa 1916 (Record Group 350-P, National Archives and Records Administration, College Park).

benevolence and colonial transformation.[45] The furniture produced at Bilibid also helped to promote the prison. The "Bilibid chair" became the best-known article produced at the prison, achieving, Jones declared, a "national, if not an international, reputation."[46] This was the kind of brand recognition that the Bureau of Education had striven, unsuccessfully, to create.

By the mid-1910s, the organization and priorities of the industrial work of the Bureau of Education and the Bureau of Prisons were more aligned than at any other point. Indeed, Jones declared the prison's industrial division to be "the biggest and one of the finest industrial schools in the Philippine Islands."[47] Its rate and quality of production, moreover, issued a "challenge to every trade school in the Philippines."[48] Certainly, Bilibid's industrial division, unlike the public schools, could devote itself full time to producing saleable articles. While Jones bragged that all 1,400 prisoners in the industrial division, including women, attended English classes for one hour a day, the other six hours were devoted to industrial work.[49] While Jones noted that it was difficult to ascertain the "permanent value of this training," he implied its success in the fact that only "two men trained in the industrial division" had ever returned to Bilibid.[50] Ignacio Villamor, the first Filipino Attorney

BUREAU OF PRISONS CATALOGUE 41

WICKER FURNITURE

CHAIRS

"The Bilibid Chair" in three sizes—small, medium, and large.

The Bilibid Chair (Nos. 1, 2, 3)

1 Small Bilibid Chair made of bamboo, heart bejuco, and black and white flat bejuco; total height, 53 inches; height of back from seat, 37 inches; diameter of seat, 21 inches........................ ₱35.00

2 Medium Bilibid Chair; total height, 59 inches; height of back from seat, 43 inches; diameter of seat, 21½ inches.. 40.00

3 Large Bilibid Chair; total height, 69 inches; height of back from seat, 51 inches; diameter of seat 23 inches.. 45.00

Crated weight, about 85 pounds; 60 cubic feet.

Figure 6.4: Bilibid chair (Catalogue of Bilibid Prison, 1927).

General, praised the benevolence of the penal system of the Philippines. In 1910, he declared that it had adopted the latest reforms of Europe and the United States, creating prisons focused on the reformation of "future free citizens."[51] Measured by its productivity, profit, and recidivism rates, Bilibid was the model industrial school, able to achieve much of what the industrial program in the public schools had been striving for: a productive, profitable, and fixed population, under colonial control and with visible markers of success, from its daily drilling to its eponymous chairs.

The industrial programs at Bilibid and in the public schools clearly influenced one another, sometimes directly. In a clear acknowledgment of the superiority of the work being done there, the Bureau of Education sent student apprentices and teachers to Bilibid to learn how to make bamboo and rattan furniture.[52] In return, the city schools assigned a teacher to instruct the female prisoners in fine needlework and primary subjects.[53] By 1915, moreover, the Director of Education bragged that a "very complete" accounting system had been adopted to track purchases and profits related to industrial work.[54] It is not clear whether this was the same accounting system developed for Bilibid in 1907, but as the prisons and schools were under the same branch of government, it would not be surprising.

Even as the two systems became more closely connected, however, they were split apart. In 1916, as part of the reorganization of the colonial government after the passage of the Jones Act, the Bureau of Prisons was transferred from the Bureau of Education to the Department of Justice.[55] Despite this, schools and prisons continued to focus heavily on their industrial programs. From this point on, indeed, the Bureau of Education redoubled its efforts in agricultural training, especially for non-Christians.

Agricultural Training

The Bureau of Education's focus on industrial education included a steady expansion of agricultural education, intended to encourage Filipinos to remain farmers rather than seek white-collar positions as teachers or clerks. Indeed, one of the most touted benefits of industrial training was that it would produce fewer graduates trained for white-collar positions which, critics claimed, did not exist.[56] Yet, many graduates of agricultural schools seem to have used their training to obtain those very jobs, rather than becoming small-scale independent farmers.

As with other schools, the system was bifurcated between Christians and non-Christians, and subdivided again between day schools and boarding schools. There were three types of schools that focused on agriculture: farm schools, agricultural schools, and farm settlement schools. Agricultural schools were boarding schools for boys in intermediate and high school grades. Farm schools were intermediate day schools with model farms. They were coeducational (girls were taught housekeeping and domestic science), and mostly, though not entirely, established for Christian students.[57] Finally, and exclusively for non-Christians, the Bureau of Education established farm settlement schools, day schools attached to a "settlement" on which non-Christian communities had been induced to reside.

The jewel in the agricultural crown was undoubtedly the Central Luzon Agricultural School (CLAS). Bureau officials mentioned the CLAS repeatedly as the model of what an agricultural school should be. Principal Kilmer O. Moe's description of the school as a "semimilitary organization" mirrored industrial schools for nonwhite children in the United States.[58] Seemingly influenced by Carlisle's discourse of detribalization, he bragged in 1929 (while no longer head of the school) that the CLAS had students from "forty different tribes," all brought together in one community where the "common denominator" was the "Anglo-Saxon tongue and the Anglo-Saxon social pattern."[59] While still the principal, moreover, he declared that the school was a "community where vagrancy is a crime."[60] Depicting the refusal to work as a crime echoed the criminalization of vagrancy as a tool of racial control in the U.S. South and imperial control in the Philippines.[61]

There were also clear links between the CLAS and the colonial penal system. In both systems, days adhered to a strict routine, labor was the basic requirement of the community, and vagrancy was a crime. In addition, both institutions divided their inmates into different classes, allowing them to earn privileges by moving up in status, and expecting the higher classes to police and discipline the lower.[62] As Michael Salman has argued, limited democracy could become "part of a disciplinary apparatus, alongside and within institutions of punishment, paternalism, instruction, and labor."[63] Allowing students and prisoners to self-regulate provided a way to efficiently control labor with a minimum of administrative oversight. Indeed, the CLAS was viewed as such an ideal disciplinary institution that the Manila Municipal Board suggested taking it as a model, instead of Iwahig, for a proposed reform school.[64]

Despite the explicitly stated purpose of agricultural and farm schools, it is questionable whether graduates of these institutions actually went on to become farmers. Many graduates did not have the funds necessary to begin farming.[65] Others seem to have viewed their training as a stepping stone to advanced study and white-collar work.[66] The CLAS did, however, become a primary exporter of agricultural teachers, particularly to the Department of Mindanao and Sulu, which by 1915 was back under the direct supervision of the Bureau of Education.[67] In 1916, Marquardt noted that the Bureau had sent 323 Filipino teachers from Luzon and the Visayas to teach there, over half of whom were graduates of the CLAS.[68] Moe also reported that from 1914 to 1916, the department had offered jobs to "almost the entire student body" because they were "not afraid to tackle a difficult job away back in the interior of the Moro and pagan country." These CLAS alumni were depicted as "imbued with pioneer spirit," starkly similar rhetoric to that of the early American teachers. [69] One factor motivating this pioneering bravery was the requirement that CLAS alumni perform one year of satisfactory service as a teacher before they were issued a certificate confirming their graduation.[70] This rule also meant that agricultural schools were geared toward producing teachers rather than farmers. Filipino graduates at times also echoed early American teachers in describing the peoples they were sent to instruct and uplift. Esteban S. Abellera wrote to Moe of his struggles in the Maganui Farm School, describing Moros as lazy and the local Datu as "bone-headed and stubborn at the same time." Abellera noted that he had learned the local dialect, however, and was working hard to win over the community, concluding, "I hope that my work will be successful in order that my certificate would be given."[71] The Bureau of Education used CLAS graduates to spread colonial education, especially to previously unreached non-Christian populations, and through that to extend the reach and power of the colonial state.

The school system also sent its alumni to work in the penal system. By the 1960s, many graduates of the University of the Philippines College of Agriculture had found employment at Iwahig.[72] Just as Bilibid and the Manila public schools exchanged instruction, and CLAS students were sent to open settlement farm schools, UP graduates could be sent to oversee the agricultural success of the Philippines' model penal colony.

From the early years of U.S. colonization, Mindanao was an incubator of farm settlements for non-Christians.[73] The primary purpose of these schools was to fix non-Christian populations in place and to encourage farming. The schools would teach the "rudiments of English, farm work, and certain household industries."[74] The students were then expected to encourage their parents to adopt the same methods and tools taught in the schools. While these schools became the model for all non-Christians, especially those previously neglected by the Bureau of Education, they heavily predominated in Agusan and Bukidnon.[75] Colonial officials argued that settlement farm schools would help teach non-Christians to become small-scale farmers. Yet in Bukidnon, where many of these schools were located, most students

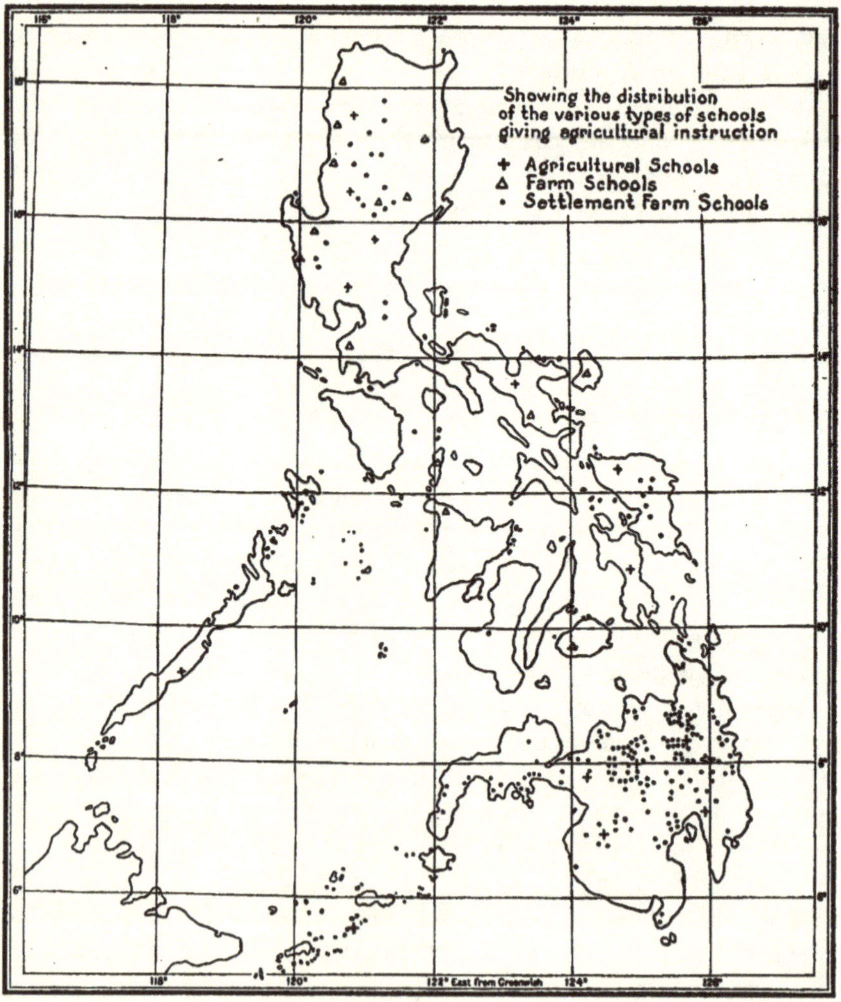

Figure 6.5: Map showing the locations of schools giving agricultural instruction, 1925 (Reports of the Director of Education 1923–1925, volume 628, RG 350, National Archives and Records Administration).

became workers on cattle ranches belonging to Americans and Filipinos from the North.[76]

The education of non-Christians became a particular focus after 1916 because of the passage of the Jones Act.[77] Maximo Kalaw noted that a primary argument against independence was the "so-called non-Christian problem." He declared, however, that unlike "the backward peoples in other parts of the globe," non-Christians in the Philippines had always been treated humanely by Americans and "their Christian brothers."[78] To further integrate non-Christians into the body politic, the Philippine Legislature expanded the settlement farm schools, especially in the Mountain Province, Nueva Vizcaya, and Mindanao-Sulu.[79] As Americans had done earlier in the century, Filipino nationalist politicians wanted to demonstrate that non-Christians were under the benevolent governance of the Philippine state.

Iwahig also held out the promise of freedom. Jones depicted the transfer of prisoners from Bilibid to Iwahig as a "graduation," where good conduct earned a "promotion from the trade school to the school farm, from industrial tutelage to industrial freedom, from the course in citizenship to a society of citizens."[80] In this valuation, to be an inmate on a penal farm was to be practically free. Beyond this, colonial officials envisioned a system in which prisoners of good standing could homestead on Palawan after their release, becoming freeholding farmers. Yet, similar to the experience of graduates of agricultural programs, Iwahig also failed to fully live up to its promises. In the early 1960s, in the midst of a glowing assessment of the penal colony, sociologist Chester Hunt noted that while the idea of homesteading had "never been abandoned," that it had not "made much progress. Like other land colonization schemes it is attractive in theory but difficult to put in practice."[81] Despite the promise of full citizenship and freedom through land ownership held out by the prisons and schools, then, it appears that those opportunities remained elusive.

Conclusion

In the mid-1920s, the Bureau of Education shifted its approach to industrial education, driven largely by the Monroe Report. In 1925, Paul Monroe of Columbia University came to the Philippines as the head of a team of researchers to evaluate the school system of the Philippines. While the Monroe Report called industrial instruction one of the "great achievements" of the Philippine schools, it criticized the focus on commercialized handicrafts, argued students could not make a living making baskets, and advocated for putting educational goals at the forefront of industrial training.[82] In response to this report, the Bureau of Education rebranded industrial education as vocational education, stressing student preparation for future careers, and the learning of skills useful for home life.[83] The Philippine Legislature and Bureau of Education devoted large amounts of money and resources to building up a Division of Vocational Education.[84]

Nationalist Filipinos like Camilo Osias, the president of National University and resident commissioner to the U.S. House of Representatives, also embraced vocational education as a tool for developing "intelligent and productive citizenship" and raising the profile of the Philippines on a global stage. Noting that the vocational work was a "distinctive feature" of the schools, Osias argued that other nations had looked to the Philippines as a model, including the United States.[85] Vocational education, then, had been absorbed into nationalist arguments for independence, proof that the islands were not only ready to manage their own affairs, but were looked to as a pedagogical leader and innovator. Praising colonial institutions was one way in which nationalists argued that the islands had achieved the benchmarks of civilization set by the United States and were ready for full sovereignty. After independence, these institutions continued to be a way for national leaders to make claims about an enlightened nation whose modern institutions were creating productive citizens.

While the Bureau of Education shifted its approach to industrial training in the late 1920s, Bilibid continued largely along the same lines. In 1927, the Bureau of Prisons published a catalogue that, alongside extensive lists of products and shipping rates, touted its "preparation of inmates for useful citizenship and honorable position in the community upon their release."[86] Bilibid's industrial division continued to focus on production and commercial profit because that was one of its primary measures of success, along with its recidivism rates. Iwahig, too, carried on with little change.

In his 1925 report, Paul Monroe declared that the "purpose of all education is to direct and control conduct."[87] This assertion was especially apt for U.S. industrial education in the schools and penal system. The overall concern was one of control—how to pacify and reform nonwhite populations perceived to be obstacles to U.S. growth and power. The U.S. colonial state attempted this by combining the tools of suasion and coercion, by offering limited self-government and promises of opportunity, while threatening punishment for those who refused to fall in line. A central aim of that control was productivity—how to turn students, non-Christians, and prisoners into productive workers for the colonial state. With similar ends, it is not surprising that schools and prisons at times utilized similar means, and learned and borrowed from one another. Industrial training in the schools and penal system was an effective means to demonstrate U.S. benevolence, shape economic development, create pliant labor, and bring the islands more closely into the U.S. imperial and industrial fold.

Notes

1. O. Garfield Jones, "Bilibid Industrial School," *Philippine Craftsman* 3, no. 2 (1914): 69–70.

2. Elisabeth M. Eittreim, *Teaching Empire: Native Americans, Filipinos, and US Imperial Education, 1879–1918* (Lawrence: University Press of Kansas, 2019); Karine Walther, "'The Same Blood as We in America': Industrial Schooling and American Empire," in *Religion and US Empire: Critical New Histories*, ed. Tisa Wenger and Sylvester A. Johnson (New York: New York University Press, 2022), 151–178; and Douglas A. Blackmon, *Slavery by Another Name: The Re-Enslavement of Black Americans from the Civil War to World War II* (New York: Anchor, 2008).

3. Aaron Abel T. Mallari, "The Bilibid Prison as an American Colonial Project in the Philippines," in "Sociology of Deviance," ed. Filomin Candaliza-Gutierrez, *Philippine Sociological Review* 60 (2012): 165–192; Michael Salman, "'Nothing Without Labor': Penology, Discipline and Independence in the Philippines Under United States Rule," in *Discrepant Histories: Translocal Essays on Filipino Cultures*, ed. Vicente L. Rafael (Philadelphia: Temple University Press, 1995), 113–129; Michael Salman, "'The Prison That Makes Men Free': The Iwahig Penal Colony and the Simulacra of the American State in the Philippines," in *Colonial Crucible: Empire in the Making of the Modern American State*, ed. Alfred W. McCoy and Francisco A. Scarano (Madison: University of Wisconsin Press, 2009), 116–128; Benjamin D. Weber, "Fearing the Flood: Transportation as Counterinsurgency in the US-Occupied Philippines," in "Transportation, Deportation and Exile: Perspectives from the Colonies in the Nineteenth and Twentieth Centuries," ed. Christian G. De Vito, Clare Anderson, and Ulbe Bosma, special issue, *International Review of Social History* 63, s26 (2018): 191–210.

4. For the teaching and commodification of embroidery in colonial schools and prisons, see Genevieve Alva Clutario, *Beauty Regimes: A History of Power and Modern Empire in the Philippines, 1898–1941* (Durham: Duke University Press, 2023).

5. Sarah Steinbock-Pratt, *Educating the Empire: American Teachers and Contested Colonization in the Philippines* (New York: Cambridge University Press, 2019); Servando D. Halili Jr., *Iconography of the New Empire: Race and Gender Images and the American Colonization of the Philippines* (Quezon City: University of the Philippines Press, 2006); Julian Go, *American Empire and the Politics of Meaning: Elite Political Cultures in the Philippines and Puerto Rico during U.S. Colonialism* (Durham: Duke University Press, 2008); and Paul A. Kramer, *The Blood of Government: Race, Empire, the United States, and the Philippines* (Chapel Hill: University of North Carolina Press, 2006).

6. Sarah Steinbock-Pratt, *Educating the Empire: American Teachers and Contested Colonization in the Philippines* (New York: Cambridge University Press, 2019).

7. Following the lead of other scholars, I will use the terms "Igorot," "Moro," and "Negrito," while acknowledging their fraught colonial history. See Rebecca Tinio McKenna, *American Imperial Pastoral: The Architecture of US Colonialism in the Philippines* (Chicago: University of Chicago Press, 2017), 8; Oliver Charbonneau, *Civilizational Imperatives: Americans, Moros, and the Colonial World (Ithaca, NY: Cornell University Press, 2020)*, xv–xvi; Frederick C. Delfin, "The Population History of the Philippines: A Genetic Overview," *Philippine Studies: Historical and Ethnographic Viewpoints* 63, no. 4 (2015): 458–459; Phillip Endicott, "Introduction: Revisiting the "Negrito" Hypothesis: A Transdisciplinary Approach to Human Prehistory in Southeast Asia," in "Revisiting the 'Negrito' Hypothesis," special issue, *Human Biology* 85, no. 1 (2013): 7–20.

8. Anne Paulet, "To Change the World: The Use of American Indian Education in the Philippines," *History of Education Quarterly* 47, no. 2 (2007): 173–202; Eittreim, *Teaching Empire*; Charbonneau, *Civilizational Imperatives*; Kilmer O. Moe, Jr., "The Influence of the Hilo Boarding School on Agricultural Education in the Pacific" (master's thesis, University of Hawaii, 1953), 40.

9. Oliver Charbonneau, "Colonizing Workers: Labor, Race, and U.S. Military Governance in the Southern Philippines," *Modern American History* 4, no. 1 (2021): 26; C. Kalani Beyer, "The Connection of Samuel Chapman Armstrong as Both Borrower and Architect of Education in Hawai'i," *History of Education Quarterly* 47, no. 1 (2007): 23–48; Sarah D. Manekin, "Spreading the Empire of Free Education, 1865–1905" (PhD diss., University of Pennsylvania, 2009).

10. Paul A. Kramer, "Transits of Race: Empire and Difference in Philippine-American Colonial History," in *Racism in the Modern World: Historical Perspectives on Cultural Transfer and Adaptation*, ed. Manfred Berg and Simon Wendt (New York: Berghahn Books, 2011), 165; Adrianne Marie Francisco, "From Subjects to Citizens: American Colonial Education and Philippine Nation-Making, 1900–1934" (PhD diss., University of California, Berkeley, 2015), 34; Glenn Anthony May, *A Past Updated: Further Essays on Philippine History and Historiography* (Quezon City: New Day Publishers, 2013), 114.

11. Frederick Fox, "Philippine Vocational Education: 1860–1898," *Philippine Studies* 24, no. 3 (1976): 264, 274, 280–282.

12. Mallari, "The Bilibid Prison," 173–174; and Charbonneau, *Civilizational Imperatives*, 115–117.

13. Salman, "'Nothing Without Labor,'" 119–120; Walther, "The Same Blood as We in America," 118.

14. Salman, "'Nothing Without Labor,'" 120.

15. Glenn Anthony May, *Social Engineering in the Philippines: The Aims, Execution, and Impact of American Colonial Policy, 1900–1913* (Westport: Greenwood Press, 1980), 89–93.

16. For more on Indian schools, see David Wallace Adams, *Education for Extinction: American Indians and the Boarding School Experience, 1875–1928* (Lawrence: University Press of Kansas, 1995), 23–27.

17. David Prescott Barrows, "Report of the General Superintendent of Education, September 15, 1904," Exhibit A, in *Reports of the Director of Education, 1899–1907*, 769. For more on Barrows's pedagogical approach, see May, *Social Engineering*, 98; Kenton J. Clymer, "Humanitarian Imperialism: David Prescott Barrows and the White Man's Burden in the Philippines," *Pacific Historical Review* 45, no. 4 (1976).

18. May, *Social Engineering*, 104; and David Prescott Barrows, "Report of the Superintendent of Education, Sept 15, 1905," Exhibit A, "Fourth Annual Report of the Secretary of Public Instruction," in *Reports of the Director of Education, 1899–1907*, 842.

19. David Prescott Barrows, Circular, No. 110, s. 1907, November 30, 1907, in "General Circulars, 1903–1909," bound manuscript, Herman B. Wells Library, Indiana University.

20. However, some Moro children in Mindanao did attend public schools with Christians, and Moro schools did roughly follow the course of study prescribed for Christian children. Najeeb M. Saleeby, "Report of the Superintendent of Schools for the Moro Province," in David Prescott Barrows, "Report of the Superintendent of Education, Sept 15, 1905, Exhibit A, Fourth Annual Report of the Secretary of Public Instruction," in *Reports of the Director of Education, 1899–1907*, 1012.

21. Fred W. Atkinson, "Report of the General Superintendent of Education for the Year Ending September 1, 1902," in *Reports of the Director of Education, 1899–1907*, 343–349; David Prescott Barrows, "Eighth Annual Report of the Director of Education, July 1, 1907, to June 30, 1908," in *Reports of the Director of Education, 1908–1914*, 71, Vol. 608, NARA; Frank R. White, "Eleventh Annual Report of the Director of Education, July 1, 1910, to June 30, 1911," 446–447; W. W. Marquardt, *Seventeenth Annual Report of the Director of Education, 1916* (Manila: Bureau of Printing, 1917), 45.

22. Shaunna Harrington, "The U.S. Colonial Education Project: Race, Citizenship & Schools in the Philippines, 1901–1916" (PhD diss., Northeastern University, 2019), 30.

23. Barrows, "Eighth Annual Report of the Director of Education," 61–62; Frank L. Crone, "Fifteenth Annual Report of the Director of Education, January 1, 1914, to December 31, 1914," in *Reports of the Director of Education, 1908–1914*, 960, Vol. 608, NARA; Harrington, "The U.S. Colonial Education Project," 101–105; Charbonneau, *Civilizational Imperatives*, 74; and Oliver Charbonneau, "'A New West in Mindanao': Settler Fantasies on the U.S. Imperial Fringe," *Journal of the Gilded Age and Progressive Era* 18, no. 3 (2019): 309.

24. Peter J. Tarr, "The Education of the Thomasites: American School Teachers in Philippine Colonial Society, 1901–1913" (PhD diss., Cornell University, 2006), 447–450.

25. White, Circular No. 135, s. 1909, December 17, 1909, "General Circulars, 1903–1909."

26. White, "Tenth Annual Report of the Director of Education for the Philippine Islands, July 1, 1909, to June 30, 1910," in *Reports of the Director of Education, 1908–1914*, 265, vol. 608, NARA.

27. White, Circular No. 135, s. 1909, December 17, 1909, in "General Circulars, 1903–1909."

28. May, *Social Engineering*, 116–117.

29. Glenn Anthony May, "The Business of Education in the Colonial Philippines, 1909–30," in *Colonial Crucible: Empire in the Making of the Modern American State*, ed. Alfred W. McCoy and Francisco A. Scarano (Madison: University of Wisconsin Press, 2009), 115–116, 119.

30. White, "Tenth Annual Report of the Director of Education for the Philippine Islands, July 1, 1909, to June 30, 1910," in *Reports of the Director of Education, 1908–1914*, 274–278.

31. "Editorial," *Philippine Craftsman* 1, no. 1 (July 1912): 69.

32. Bureau of Education Philippines, *The Philippine Public Schools at the Panama-Pacific International Exposition* (San Francisco: Press of Marnell & Company, 1915), 1.

33. Frank L. Crone, "Sixteenth Annual Report of the Director of Education, January 1, 1915, to December 31, 1915," in *Reports of the Director of Education, 1912–1915*, 35, vol. 629, NARA; and Marquardt, *Seventeenth Annual Report of the Director of Education*, 1916, 11.

34. Marquardt, *Eighteenth Annual Report of the Director of Education, 1917* (Manila: Bureau of Printing, 1918), 43–45. This work had been handled by a Sales Agency from 1912 to 1915. Crone, "Fifteenth Annual Report of the Director of Education, January 1, 1914, to December 31, 1914," in *Reports of the Director of Education, 1912–1915*, 1001, vol. 629, NARA.

35. Marquardt, *Eighteenth Annual Report of the Director of Education, 1917*, 46.

36. Mallari, "The Bilibid Prison," 175.

37. Mallari, "The Bilibid Prison," 178.

38. Jones, "Bilibid Industrial School," 90–91.

39. Prisoners sentenced to less than six months, or those with a poor conduct rating, were mostly put to work on road-building projects. Initially, the industrial division paid the prison division for the labor of the prisoners. In 1916, Bilibid began paying prisoners who worked in the industrial shops, with wages determined by work and skill rating. See Salman, "'Nothing Without Labor,'" 116–117; Jones, "Bilibid Industrial School," 73–75, 85; and Mallari, "The Bilibid Prison," 177–178.

40. Jones, "Bilibid Industrial School," 74–75. See also George N. Wolfe, *Annual Report of the Bureau of Prisons for the Fiscal Year 1908* (Manila: Bureau of Printing, 1909), 30, in Worcester Philippine History Collection, University of Michigan.

41. Wolfe, *Annual Report of the Bureau of Prisons for the Fiscal Year 1908*, 30.

42. Student and parent opposition could delay or disrupt industrial education. Luther Bewley noted in 1925 that opposition in two non-Christian divisions to farmwork in the schools led the bureau to transition from agricultural to barrio schools for the children there. Luther Bewley, "Twenty-Sixth Annual Report of the Director of Education, 1925," in *Reports of the Director of Education 1923–1925*, 64, vol 628, NARA. For more on opposition to industrial education, see May, *Social Engineering*, 93; and Douglas Foley, "Colonialism and Schooling in the Philippines from 1898 to 1970," in *Education and Colonialism*, ed. Philip G. Altbach and Gail Paradise Kelly (New York: Longman Inc., 1978), 81.

43. W. Morgan Shuster, "Report of the Secretary for Public Instruction," in *Eighth Annual Report of the Philippine Commission to the Secretary of War*, 1907, Part III (Washington, D.C.: Government Printing Office: 1908), 172–173.

44. Mallari, "The Bilibid Prison," 177.

45. *Catalogue of Products of the Industrial Division of Bilibid Prison* (Manila: Government of the Philippine Islands, 1927), 19; and Mallari, "The Bilibid Prison," 184. This kind of penal tourism was not limited to the colonial Philippines. See Jacqueline Z. Wilson et al., eds., *The Palgrave Handbook of Prison Tourism* (London: Palgrave Macmillan, 2017), 3.

46. Jones, "Bilibid Industrial School," 79. The Bilibid chair went on to have a life of its own, used in photo shoots for presidents and Black Panthers. Mario Alvaro Limos, "How a Rattan Chair from Bilibid Became the World's Seat of Power," *Esquire*, June 24, 2020, www.esquiremag.ph/long-reads/features/rattan-peacock-chair-bilibid-a00293-20200624-lfrm.

47. Jones, "Bilibid Industrial School," 69.

48. Jones, "Bilibid Industrial School," 92.

49. Jones, "Bilibid Industrial School," 90–91.

50. Jones, "Bilibid Industrial School," 93.

51. Salman, "'Nothing Without Labor,'" 113.

52. Jones, "Bilibid Industrial School," 79–81; and Andrew W. Cain, "The Ninth Vacation Assembly," *Philippine Craftsman* 3, no. 1 (July 1914): 34.

53. Marquardt, *Seventeenth Annual Report of the Director of Education*, 1916, 40. See also Clutario, *Beauty Regimes*, 139–182.

54. Crone, "Sixteenth Annual Report of the Director of Education, January 1, 1915, to December 31, 1915," in *Reports of the Director of Education, 1912–1915*, 39.

55. Francis Burton Harrison, *Report of the Governor General of the Philippine Islands to the Secretary of War, 1916* (Washington, D.C.: Government Printing Office, 1917), 4.

56. See, for example, Barrows, "Report of the General Superintendent of Education, Sept. 1, 1902 to Sept 30, 1903," in *Reports of the Director of Education, 1899–1907*, 471–472.

57. "What the Philippine Public Schools Are Doing in Agriculture," *Philippine Craftsman* 2, no. 8 (1914): 539–540; and Fry, *A History of the Mountain Province*, 221–222.

58. Moe, "The Central Luzon Agricultural School," *Philippine Craftsman* 2, no. 8 (February 1914): 595.

59. Moe, Jr., "The Influence of the Hilo Boarding School," 40.

60. Moe, "The Central Luzon Agricultural School," 590.

61. See Blackmon, *Slavery by Another Name*; and Steinbock-Pratt, *Educating the Empire*, 152.

62. Jones, "Bilibid Industrial School," 73; Moe, "The Central Luzon Agricultural School," 595–596; "Police (All Colonists), Iwahig Penal Colony," "Bureau of Prisons," photograph book, compiled by Frances Burton Harrison, Library of Congress; Salman, "'The Prison That Makes Men Free,'" 120–124; Salman, "'Nothing Without Labor,'" 120; and Wolfe, *Annual Report of the Bureau of Prisons for the Fiscal Year 1908*, 58;

63. Salman, "'The Prison That Makes Men Free,'" 126.

64. "Munoz Plan for Reform School," *The Cablenews-American*, 1918, 3.

65. Bewley, "Twenty-Sixth Annual Report."

66. Joseph A. Cocannouer, "The Indang Farm School," *Philippine Craftsman* 2, no. 8 (1914): 579; and Moe, "The Central Luzon Agricultural School," 600.

67. Harrington, "The U.S. Colonial Education Project," 167; and Charbonneau, *Civilizational Imperatives*, 17–18. As Charbonneau notes, this period inaugurated "Filipinization" in Mindanao-Sulu, in which American officials were steadily replaced by Christian Filipinos.

68. Marquardt, *Seventeenth Annual Report of the Director of Education, 1916*, 47.

69. Luther Bewley, "Twenty-Second Annual Report of the Director of Education, Jan 1, to Dec 31, 1921," in *Reports of the Director of Education 1921–1923*, 40, vol. 610, NARA.

70. Moe, "Results of Agricultural Education," 250–251.

71. Esteban S. Abellera, Letter to Kilmer O. Moe, July 1, 1915, vol. 3, box 5, W. W. Marquardt Papers, Bentley Library, University of Michigan, Ann Arbor; for more on American teachers' depictions of Filipinos, see Steinbock-Pratt, *Educating the Empire*.

72. Chester L. Hunt, "Iwahig Penal Colony: Freedom as a Social Therapy," *Philippine Sociological Review* 9, no. 1/2 (1961), 36.

73. Leonard Woods, *Annual Report of the Governor of the Moro Province, September 1, 1903 to August 31, 1904* (Washington, D.C.: Government Printing Office, 1904), 9; Saleeby, "Report of the Superintendent of Schools for the Moro Province," 1014; Barrows, "Eighth Annual Report of the Director of Education," 121; and Charbonneau, "Colonizing Workers," 41–42.

74. "What the Philippine Public Schools Are Doing in Agriculture," 541.

75. "What the Philippine Public Schools Are Doing in Agriculture," 541.

76. Ronald K. Edgerton and Becky Richards Edgerton, "Growing Up in a Frontier Town: The Bukidnon Experience in Malaybalay," *Philippine Quarterly of Culture and Society* 10, no. 3 (1982): 126–128.

77. Fry, *A History of the Mountain Province*, 107–108.

78. Maximo M. Kalaw, "Recent Policy Towards the Non-Christian People of the Philippines," *Journal of International Relations* 10, no. 1 (1919): 1.

79. Marquardt, *Seventeenth Annual Report of the Director of Education, 1916*, 44–45.

80. Jones, "Bilibid Industrial School," 74.

81. Hunt, "Iwahig Penal Colony," 38–41.

82. The Board of Education Survey, *A Survey of the Educational System of the Philippine Islands* (Manila: Bureau of Printing, 1925), 59–61, 281–282, 288, HaithiTrust, https://hdl.handle.net/2027/miun.ahk8495.0001.001.

83. Bewley, "Twenty-Seventh Annual Report of the Director of Education, 1926," in *Reports of the Director of Education 1925–1929*, 48–49, vol 613, NARA; and Gilbert S. Perez, "Vitalizing the Work in the Elementary School Shop," *Philippine Public Schools* 1, no. 2 (1928): 61–62.

84. Cecilio Putong, "A Proposal for the Administration of Vocational Agricultural Education in the Philippine Islands," *Philippine Public Schools* 2, no. 2 (1929): 51; and Bewley, "Twenty-Ninth Annual Report of the Director of Education, 1928," in *Reports of the Director of Education 1925–1929*, 9, vol 613, NARA.

85. Camilo Osias, "The Value of Vocational Education," *Philippine Public Schools* 1, no. 4–5 (1928): 176–180.

86. *Catalogue of Products of the Industrial Division of Bilibid Prison*, 22.

87. Board of Education Survey, *A Survey of the Educational System of the Philippine Islands*, 94.

CHAPTER 7

"Training the Heart, the Head, and the Hand": Colonial Education, Missionaries, and Industrial Schools in Korea, 1906–1930

Dolf-Alexander Neuhaus

Introduction

Vocational and industrial training loomed large in Japanese colonial education in Korea. Between 1910 and 1945, Japanese authorities implemented an education system that concentrated on industrial education for Koreans, who were generally portrayed as possessing a low level of civilization (J: *mindo*).[1] Geared toward assimilating Koreans as Japanese subjects, this system was at the same time characterized by pervasive segregation and discrimination with severely restricted access to higher education for Koreans. After annexation in 1910, the colonial education system was superimposed upon a fragmented, yet evolving structure consisting of traditional and modern schools as well as foreign missionary schools that continued to operate throughout most of the colonial period.[2] Promoting manual labor, industrial education constituted a salient feature of the proclaimed Japanese civilizing mission as it was deemed befitting to "lazy" Koreans by Westerners and Japanese alike.[3] For the Japanese colonial power, civilizing Koreans through labor education furthermore fulfilled the empire's economic needs while developing a sufficiently educated, yet subordinate colonial population.

The nurturing of a cheap colonial workforce by means of industrial education was certainly not unique to the Japanese Empire.[4] One of the idiosyncrasies of the Korean case, however, is the long-standing presence of missionaries from Europe and North America in a colony ruled by a non-Western power.

Especially during the early phase of colonial rule, Japanese authorities saw themselves confronted with a competing transfer route for the reception of civilization and modernity due to the presence of missionaries. Utilizing the social gospel, missionaries of the Young Men's Christian Association (YMCA) encouraged industrial and agrarian development in large parts of Asia, so much so that the missionaries' medical, educational, and social work became a constitutive element of Protestant Christianity in Korea.[5] While missionaries in Korea functioned as bearers of America-centric capitalist modernity, mission schools concurrently provided Koreans with a viable alternative to Japanese colonial modernity, considerably complicating the colonizer's assimilation strategy.[6]

Nevertheless, scholarship has often neglected industrial education in colonial Korea.[7] This is surprising given the abundance of excellent works on colonial education as well as the history of missionaries and their impact on Korean education.[8] This chapter sets out to fill this scholarly gap by exploring the interactions around industrial education between Japanese colonial authorities, Western missionaries, and Korean Protestants. Delving into a broad variety of sources, including Korean writings, missionary reports and journals, such as the *Korea Mission Field*,[9] and official documents, I argue that the efforts for industrial education promoted by colonial authorities and Western missionaries frequently intersected and overlapped, yet on occasion also collided with one another. The latter was invariably the case when Koreans became an active part of the equation.[10] Research has repeatedly demonstrated that local agency mattered in the history of missions and imperialism.[11] While Western missionaries and the YMCA strove to teach Korean pupils about industrial labor and capitalism through "Industrial Education Departments" and instill in them a Protestant work ethic, Korean nationalist leaders, many of whom had converted to Christianity, saw industrialization as a requirement to attain "civilization and enlightenment" and "self-strengthening" in order to preserve independence. This call for the development of an industry-based capitalist economy in Korea, which they saw as a prerequisite to becoming modern, predated the colonial period.[12] Their activities, at least on the surface, nevertheless dovetailed neatly with the overall educational strategy of Japanese authorities in Korea, which stressed vocational education. As in many cases, colonization was a struggle over consciousness in Korea, too. Missionaries could sometimes act as valuable agents of imperial expansion, which occurred less at the articulated "ideological" level, but rather at the unconscious "hegemonic" level.[13] Yet, missionaries also stood as "central transmission belts" at the intersections of numerous activities,[14] whose intensive world-spanning interconnections favored the emergence of an imagined global community providing a refuge for Koreans under Japanese rule.

The following provides a comprehensive overview of government and missionary industrial education policies and institutions in Korea during the early twentieth century. After outlining the early development of modern education in Korea, the chapter analyzes industrial education as a Japanese approach to assimilation, and parses the relationship between Western missionaries and Korean reformers, who, at different times, both engaged with and opposed colonial governance.

The Modernization of Korean Education Prior to 1910

At the end of the nineteenth century, Chosŏn-Korea experienced a profound transformation affecting all aspects of society and politics. Internal upheavals, including the persecution of Catholics during the 1860s, violent power struggles in the early 1880s, and the Tonghak Rebellion, were compounded by foreign infringements on Korean sovereignty. Japan's gunboat diplomacy forced the Chosŏn-court to sign the unequal Kanghwa Treaty in 1876, and Japan subsequently vied with China and Czarist Russia for supremacy over Korea during the First Sino-Japanese War and the Russo-Japanese War.[15] The ensuing rapid sociopolitical transformation also considerably affected the field of education. Until its formal abolition in the wake of the modernizing Kabo reforms in 1895, the constitutive element of Chosŏn-Korean society had been the caste system that effectively impeded any kind of social mobility. The literati class (*yangban*) at the top frowned upon manual labor done by members of the lower castes, outcasts and slaves (K: *nobi*), even pertaining to members of the *chungin* caste that formed the upper middle class of hereditary specialist professionals, including technicians, medical practitioners, and interpreters.[16] Hence, traditional education in Korea concentrated on the study of the Chinese classics, mainly of the neo-Confucian school, and was monopolized by the members of the *yangban* class. Thorough knowledge of these classic works was a prerequisite for successfully passing different levels of the civil examination system (K: *kwagŏ*), which in turn was a requirement to hold public office. Reiterating commonly held Western stereotypes of the era, British travel writer Isabella Bird Bishop concluded in 1898 that the traditional Korean education system only produced such unfavorable traits as "narrowness, grooviness, conceit, superciliousness, a false pride which despises manual labor, a selfish individualism, destructive of generous public spirit and social trustfulness."[17] This system was finally scrapped in the sweeping Kabo reforms to dismantle the Confucian grip on the education system.

The landscape of the Korean education system began to diversify at the turn of the twentieth century. Notably, Kojong, the penultimate ruler of the Korean Chosŏn dynasty, opened his country to Christian missionaries in 1884, leading to an influx of North American missionaries who established numerous schools across the country. Apart from Bible studies, syllabi comprised subjects such as world history, geography, mathematics, natural sciences, English, physical education, and industrial crafts. Presbyterian and Methodist missionaries, the most prominent foreign denominations in Korea, concentrated on improving the educational and medical systems as an effective approach to evangelization in Korea. Additionally, according to the Nevius plan, the control and maintenance of churches were largely left to the Korean converts, considerably facilitating the success of Protestantism in modern Korea and the strengthening of nascent Korean nationalism.[18] The emphasis of missionary schools on caring for the poor, promoting equality before God, and providing basic education for socially marginalized groups, including women, significantly boosted their popularity.[19]

A new type of Korean school, such as Wŏnsan haksa (Wŏnsan Academy), offered traditional Chinese classics in combination with subjects such as mathematics, physics, mechanics, agriculture, sericulture, mining, law, and geography, breaking with the traditional curriculum of Confucian education.[20] Nevertheless, schools like the Pai Chai Academy (Paechae haktang) for boys mainly catered to the elite to foster capable bureaucrats to help Korea's self-strengthening efforts. Founded in 1885 by Methodist missionary Henry G. Appenzeller, the academy was one of the first institutions in the country that employed pupils in industrial education departments.[21] Elevated to college status in 1895, and partly funded by the Korean government, the school was described by Isabella Bird Bishop in 1898 as having "a flourishing industrial department, which includes a tri-lingual [sic] press and a book-binding establishment, both of which have full employment."[22] Rather than providing industrial training in preparation for professional life, however, these early industrial departments were intended to aid pupils with the financial means necessary for their education.[23] Still, the school's industrial department had to close down some time later due to increased competition by professional presses and publishing companies.[24]

In addition to the efforts of missionaries and the state, private initiatives also became active in the field of education. Leading members of the Independence Club (K: *Tongnip hyŏphoe*), including Western-educated reformers such as Sŏ Chaep'il (Philipp Jaisohn) and Yun Ch'iho, blisteringly criticized Confucian education while advocating enlightenment and self-strengthening through their bilingual newspaper *The Independent* (K: *Tongnip sinmun*).[25] Although the Independence Club existed for only a brief period between April 1896 and December 1898, its long-term significance for the further course of Korean history cannot be underestimated. Among its members were influential reformers and future independence activists and presidents. Moreover, many former members of the Independence Club such as Yun Ch'iho later occupied important positions in the leadership of the Korean YMCA, so much so that the latter organization resembled a revitalized version of the former.[26] Along with a Student Association at Pai Chai College, the official Korean branch of the YMCA was launched in Seoul in October 1903 as the Hwangsŏng YMCA (Seoul YMCA), which within a few months boasted a membership of 263 people.[27] More YMCAs were soon inaugurated in other cities of the country, and the national association became a member of the World Alliance in 1905.[28] A self-proclaimed objective of the Seoul YMCA was the advancement of modernization in Korea through the introduction and the spread of technical, commercial, as well as physical education in order to root out the persistent contempt for physical activities. In Seoul, the YMCA offered English and Bible classes as well as industrial and sports education.[29]

Patriotic movements also carried on the legacy of the Independence Club during the following decade.[30] Defeating Russia in 1905 had allowed Japan to establish ample control over Korean politics in the protectorate treaty. Under the protectorate, both state-level and private actors endeavored to enhance access to modern schooling and further Korean self-strengthening efforts. This gave rise to a number of Korean nationalist movements that are habitually subsumed under the moniker "patriotic enlightenment movement" (K: *Aeguk kyemong undong*).[31] Fearing the loss of sovereignty, modern education and the development of an indigenous industry were considered vital by many Korean political and societal leaders. In 1906, Chang Chiyŏn (1864–1921) and others penned the "Manifesto of the Korean Association for Self-Strengthening," which they published in the daily newspaper *Hwangsŏng sinmun* (*Imperial Capital News*). The authors emphasized the need for simultaneous development of education and industry to promote wisdom and enhance the nation's power.[32]

In a similar vein, the Sinminhoe (New People's Association) promoted "education to save the nation" with the aim of instilling Koreans with patriotic ideas.[33] Historian Pak Ŭnsik (1859–1925), one of the Sinminhoe's leaders, buttressed the absolute necessity of education in the modern world, urging his countrymen to engage in "new learning, and developing new knowledge" in order to avoid being "reduced to a lower status in the world due to a lack of knowledge and skills, thereby becoming the slaves and victims of foreigners."[34] In 1908, Pak published another well-known essay on technological innovation in which he advocated the need to survive in a capitalist world by improving and refining Korean industrial products and skilled labor.[35] While not explicitly endorsing industrial education, Pak, Chang, and other proponents of patriotic enlightenment had realized the need to improve education as a means of furthering industrial production to make effective use of the country's wealth of raw materials. This national industrial movement sought to counter the intensifying Japanese economic invasion.[36] Concomitantly, the country witnessed a wave of private and Christian school openings, which often proved very popular and helped set the intellectual foundation for the establishment of industrial schools in Korea. By 1910 there were approximately 2,000 of these institutions with an estimated 200,000 students.[37] In addition, in 1909, the Korean government, already under the influence of the Japanese, decreed the establishment of industrial schools. These, however, soon ran into financial difficulties.[38]

To self-strengthen, in the perceptions of the Korean nationalists, meant to ensure independence. This overlapped with the efforts of Western missionaries in Korea, whose schools enjoyed a high degree of popularity. On the eve of annexation, missionaries in Korea did fulfill a role as spiritual leaders, and were often perceived as powerful allies against Russia and Japan, although they officially adopted a strict policy of separating religion and politics.[39]

Industrial Education in Japanese Colonial Education Policy

After annexation in August 1910, colonial authorities prioritized vocational skills and tried to discourage Koreans from pursuing "unnecessary" higher education in order to mold loyal and acquiescent colonial subjects. While the educational activities of Korean nationalists before 1910 had borne some fruit, and a number of private institutions including local and missionary schools were permitted to keep operating under the colonial regime, Japanese authorities wasted no time in setting a new general framework that suited the needs of the colonial government, clearly advocating vocational education for Koreans. Within a year of annexation, the newly appointed Governor-General in Korea, Terauchi Masatake, issued the first Korean Educational Ordinance (J: *Chōsen kyōikurei*). The general provisions stipulated that "the essential principle of education in Chosen [Korea] shall be the making of loyal and good subjects by giving instruction on the basis of the Imperial Rescript concerning Education . . . [and] shall be adapted to the needs of the times, and the condition of the people."[40] The guiding principle and ultimate pedagogical objective was the assimilation of Koreans as sequacious imperial subjects. Therefore, the highest priorities were the "cultivation of moral character" and "diffusion of the national language" (Japanese) in accordance with the current "cultural level" (*mindo*) of the Koreans.[41] The Japanese term *mindo* signified a key concept, which, borrowed from Montesquieu, provided the justification for pervasive segregation in the colonial educational system and simultaneous assimilation of Koreans as loyal subjects.[42] The conspicuous absence of any clause pertaining to universities and colleges merely buttressed this overall strategy. In 1913, Terauchi instructed provincial governments in Korea that "people in Korea today are not quite ready for noble learning yet. We must therefore place the emphasis on giving common education on matters familiar to them to make people capable of performing their allotted work well."[43] The public common schools in Korea during Japanese rule emphasized Japanese (or "national") language instruction and morals, whereas subjects like Korean geography and history, at least prior to 1922, were not taught. Instead, education for Koreans focused on basic and vocational skills for them to perform more trivial forms of handiwork. Even higher education prioritized practical skills like agriculture and engineering.[44]

Apart from determining the general direction of colonial education policy, the Educational Ordinance laid out three different types of schools that were to form the colonial school system. Article four roughly classified three kinds of education: common (*futsū*), industrial (*jitsugyō*), and special (*senmon*). Industrial education was defined as aiming "at imparting knowledge and art concerning agriculture, commerce, technical industry, and so forth," while common education served to engender national characteristics and to spread the national language (i.e. Japanese). By the same token, "special education" was geared toward advanced-level training of specialists and experts, including lawyers, engineers, and doctors. According to the ordinance of 1911, industrial schools (*jitsugyō gakkō*) provided the necessary training for "students wishing to engage in agriculture, commerce and technical industry" for two or three years. Industrial schools included agricultural, commercial, technical, and elementary industrial schools. In order to be eligible for an industrial school, the Governor-General stipulated that students "shall be not less than twelve years of age, and graduates of a Common School," or possess equivalent or superior qualifications. The ordinance did not immediately specify the duration of study and the requirements for students at the elementary industrial school.[45] In May 1918, one account reported seventeen agricultural, three commercial, and sixty-seven elementary industrial schools in Korea. The same article further revealed that industrial education was not limited to special institutions but was also taught at common schools. Out of a total of 460 common schools, 355 offered courses in agriculture and 140 instructed pupils in handicraft.[46]

Early efforts "to stimulate the improvement of agriculture and industry among the Koreans" had already been made under the auspices of the Japanese Residency General. In 1907, an exhibition was held in Seoul, and "Local Industrial Museums" were established in 1908 in every province. In 1908 and 1909, for example, several industrial training associations received funds from the central government in Seoul, including the Taiku (Daegu) Industrial Training Association, the Tamyang (Damyang) Industrial Training Place, and the South Korea Matting Manufacturing Company in Daegu, as well as a number of similar endeavors. Most of these associations specialized in weaving, matting, and bamboo work.[47] Showcasing its own achievements, the Governor-General's office claimed that "in 1909, reform measures entered the third stage, being directed towards encouraging and stimulating industrial and technical ideas in order to meet the existing condition of the country. This was effected by promulgating Regulations concerning Industrial Schools and by increasing the curriculum of agriculture, commerce and industry in the Normal School, High School and Common Schools."[48] In common schools, "the supervision of text-books was strictly enforced," while "lessons concerning industry in supplementary course of common schools were also made compulsory." Furthermore, the possibility to add lessons relating to agriculture and commerce to the curriculum existed "according to local conditions." This seemed to address a certain shortage of qualified personnel, as the same report mentioned that graduates of the Industrial Training School and the Agricultural Dendrological School were employed as instructors in the common schools.[49]

On the surface, Japanese officials shared with Korean nationalists the view that general ignorance about science and agricultural and industrial knowledge prevailed in the country.[50] However, Japanese colonial authorities by no means supported or even approved of the activities of the Sinminhoe. While Korea was still formally an independent country before 1910, Japanese officials began to work systematically toward taking over the languishing Korean public school system by employing Japanese teachers and issuing new textbooks, among other things. Anticipating assimilation policy after annexation, this served to form a population more amenable to the Japanese.[51] From the outset, Japanese authorities perceived the "education for the nation" movement of the Sinminhoe as an obstacle to their plans of pursuing a policy of assimilation and incorporation of Korea into the Japanese Empire.[52] Shortly after the annexation, the organization was quickly forced to dissolve due to fabricated allegations of conspiracy to assassinate Governor-General Terauchi in what became known as the "Case of the 105." Given the large number of prominent Christians and high-ranking YMCA officials among its members, the crackdown raised serious concern among missionaries in Europe and North America despite Japanese efforts to reassure them that the crackdown was not anti-Christian in nature and their work was not in jeopardy.[53] At the Hugh O'Neill Jr. Industrial Academy, a large Presbyterian mission school in Sŏnch'ŏn, so many pupils and teachers were arrested and hurried to jail that the school had to temporarily close.[54] Eventually, many schools established in the course of the "education for the nation" movement, such as the Taesŏng School in Pyongyang (P'yŏngyang), were forced to close following the crackdown.[55] The "Case of the 105" ushered in a decade of severe repression of political and cultural life in Korea, which is known as the "dark period" (K: *amhŭkki*).

Subsequently, colonial authorities remained wary of the educational activities of missionaries in Korea. In 1919, the fact that many Christians participated in the March First Independence Movement led Japanese media and government officials to accuse Western missionaries of inciting their Korean congregants to revolt against their Japanese overlords. From the vantage point of Western missionaries, governmental pressure jeopardized the continuance of Christian schools due to the lack of funds that they desperately needed to meet government standards for infrastructure and equipment. Government regulation furthermore burdened public and private schools with teaching Japanese language, history, and moral classes, while graduates of missionary schools found it difficult to get into the most prestigious institutions of higher education or find attractive employment. Moreover, religious instruction was prohibited until the ban was relaxed in 1922.[56]

Nevertheless, demand for private Korean schools as well as missionary schools remained strong. In 1915, Japanese authorities in Korea legalized private technical schools as institutions of postsecondary education. Initially, only four schools received recognition under the Regulations for Technical Schools (J: *senmon gakkō kisoku*) that had originally been created by the Korean government in the early 1900s: Keijō (Seoul) Professional School of Law, Keijō Professional School of Medicine, Keijō Professional School of Industry founded in 1906, and the Agricultural and Dendrological School in Suwŏn, which had been established under the name Agricultural, Commercial, and Industrial School in 1904.[57] Simultaneously, the Government-General issued regulations concerning private education in Korea. Christian institutions such as Pai Chai College (1916) and the Anglo-Christian School (1917) obtained status as higher common schools.[58]

Although the colonial education system was revamped several times through the reissuing of updated Imperial Rescripts of Education in 1922 and 1938, its overall mission remained largely the same. The *Manual of Education in Chosen* [Chōsen] published by Japanese colonial authorities in 1920 defined industrial schools as institutions "in which students wishing to engage in agriculture, commerce or technical industry are trained, giving them the necessary education therefor [sic]."[59] The *Manual* indiscriminately applied the label "industrial school" to agricultural schools, commercial schools, technical schools, and elementary industrial schools, and, in 1920, counted seventeen agricultural schools and three commercial schools that altogether had 1,849 students and employed 108 teachers. The report specified the educational objectives of the existing sixty-seven public elementary industrial schools, which focused on eradicating "the deep-rooted and evil habit . . . of despising industry and abhorring toil and labour" and improving local agricultural industry according to the Governor-General's industrial policy.[60] The claim that Koreans despised manual labor was a recurring theme not only in official Japanese reports but also in missionary publications.[61] Colonial authorities contended that such evil had to be stamped out, preferably through industrial education, and utilized this assertion to justify the omnipresent segregation in the education system. E. W. Koons, who presented a survey of the schools of Seoul in *The Korea Magazine*, elucidated the content of teaching at the elementary industrial schools for boys in August 1918: "One of these [elementary industrial schools] is connected with the Oh-I-Dong, primary School (opposite Dr. Gale's residence at Yun Dong). It takes graduates of the Primary School and gives them a year's course in either carpentry or blacksmithing, with some classroom work, and a great deal of practice."[62] Moreover, the content of contemporary schoolbooks reveals that depictions of manual labor appeared much more often in the textbooks for Koreans. Illustrations often portrayed Koreans in subservient positions while Japanese are shown in positions of authority and moral leadership, as teachers or policemen.[63]

By self-confidently advertising an "enlightened" educational policy as part of its civilizing mission in Korea and Taiwan, the Japanese government sought to assert its position among the "civilized" nations of the world, i.e., the other imperial powers.[64] Compiled in English, the *Manual* was clearly directed at a foreign, mainly Western audience, and thus served the purpose of showcasing the progress of Japan's civilizing mission and simultaneously delegitimizing Korean traditional knowledge: "It is observed with satisfaction that graduates from industrial schools who are found in different parts in the peninsula, are proving effective instruments in securing improvements in the species of the various crops, selection of seeds, improvement in the method of cultivation and the method of preparing them for the market and are helping in livestock farming, sericulture and forestry, teaching their neighbors to discard old methods and adopt advanced ones, thereby contributing either directly or indirectly to improvement in local agricultural matters."[65]

The *Manual* furthermore acknowledged the existence of private industrial schools, among which the Zenrin Commercial School in Seoul stood out as an "excellent institution" with more than 164 students. Irrespective of official efforts to delegitimize them, private schools, including missionary schools, persevered throughout the colonial period.[66] The state even subsidized certain private industrial schools that achieved commendable outcomes.[67]

Not all foreign observers were oblivious to the defects, and perhaps even to the true objectives, of the colonial education system. In 1920, Richard H. Ritter criticized the Japanese colonial government for neglecting the Koreans who were suffering under Japanese competition. Ritter served as a missionary to China with the Princeton-in-Peking program, doing social work for the YMCA from 1920 to 1929. While praising the achievements of Japanese-led industrial education in Korea, he pointed out that "all these improvements are preponderantly to the advantage of the Japanese colonists" who were "perhaps the more advanced race." This further accentuated "the difference between the conquered race and the conquering race," whether intended or not.[68] During the early 1920s, after ten years of repressive colonial rule, Koreans were hit by economic hardships caused by forced integration into a political economy that was tailored to meet the needs and objectives of the Japanese Empire. One reason for the quick spread of the March First Independence Movement in 1919 had in fact been pent-up discontent about systematic discrimination in education and employment, as even some liberal Japanese commentators acknowledged.[69]

After 1919, military repression segued into more lenient "cultural rule" in an attempt to accommodate Korean demands for better access to education by implementing quantitative expansion of the school system. Yet, progress was slow. Compounded by the economic crisis of the 1920s primarily affecting rural communities, the education system suffered from chronic underfunding and a lack of qualified teachers. The Governor-General tried to address this issue in 1929 by amending the regulations for primary school education and introducing additional mandatory courses to strengthen vocational training during early education.[70] The new policy aimed to forge elementary school graduates into capable and efficient farmers after leaving school.[71] Knowledge-oriented education for Koreans, including science and literature, was not encouraged by authorities, and colonial education continued to focus on low-level job training during the 1930s, following a strategy that was known as practical education policy.[72] Courses comprised basic agricultural manual labor like weeding, sowing, plowing, rearing silkworms, harvesting, selling crops at the market, and building barns. The intention that lay behind this strategy was to force Koreans to remain in the agricultural sector in order to prevent the spread of socialist thought and to sway Koreans away from anti-Japanese movements.[73] By 1938, when Japan was already embroiled in an intensifying war against China that would soon escalate into the Asia-Pacific War, the Educational Ordinance was once again revamped and the separation of schools for Japanese and Koreans was abolished in name. Aside from 3,220 elementary schools hosting over a million pupils, there were only 50 middle schools with 25,822 pupils and 54 girls' high schools with 20,716 pupils throughout the country. Concomitantly, 135 elementary industrial schools taught 7,413 pupils, while only 77 industrial schools had as many as 22,331 pupils. Considering that these figures included Japanese children, access to education, particularly above the primary level, evidently remained severely restricted for Koreans.[74]

While Japan proper had seen a similar debate about adapting education to real life a few years earlier, the vocational program was implemented in Korean primary schools as "terminal" education.[75] In Japan, by contrast, it was introduced in the middle schools. Generally, the education provided to

Korean children was greatly inferior in quality and quantity to that of the Japanese settlers. The colonial education system was hence one of the most powerful tools at the disposal of the Japanese to facilitate assimilation and ensure that Koreans would remain docile colonial subjects.

Missionaries and Industrial Education in Korea

In his booklet *Modern Education in Korea*, Horace Horton Underwood elucidated the importance that Western missionaries attributed to the Industrial School of the YMCA in Seoul: "The Y. M. C. A. Industrial School in Seoul is really the only missionary school in the country having the single objective of industrial or trade training without the entangling features of student help. . . . It is now housed in a large three-story building, an annex to the main building of the Y. M. C. A. in the center of the city. After experimenting with different trades it now offers courses in printing, cabinet-making, wicker-work, machine shop work, photography and soap and candle making."[76]

Horace H. Underwood was the son of Presbyterian missionary and educator Horace Grant Underwood, who had arrived in Korea in 1884 and was later instrumental in the founding of the Seoul YMCA as well as Chosŏn Christian College, the predecessor of modern Yonsei University. The younger Underwood's view was corroborated by Richard H. Ritter's earlier observations while traveling Korea in 1920: "The only real industrial mission school is that of the Young Men's Christian Association in Seoul. Others have industrial departments, but these may without exception be reckoned as self-help departments, and not, accurately speaking, as industrial."[77]

The YMCA's Industrial School had been founded shortly after the 1906 arrival in Seoul of George A. Gregg, who continuously served as its director until 1928. Prior to 1906 there were no technical or industrial schools in Seoul aside from sporadic attempts that had failed.[78] Gregg had previously served at the Industrial Institute in Hartford, Connecticut, and as secretary of the Educational Department of the YMCA's International Committee. After his arrival in Korea, he immediately set up a temporary building equipped with various furniture, tools, and devices, which was used as a trade school offering carpentry, plumbing, and blacksmithing classes.[79] The construction of the Seoul YMCA main building was completed in 1908, and in December 1914 the Industrial Building behind the main building was completed.[80] This enabled the Seoul YMCA to shift its focus toward a stronger emphasis on industrial education in autumn 1916, when it had to eliminate general education and Korean literature classes due to revised regulations for private schools. The Industrial School assumed the functions of a school as well as a commercial cooperative.[81]

Only members of the YMCA were allowed to attend the YMCA Industrial School, and every aspiring student had to present "a letter from a responsible guarantor" upon entrance.[82] The intent was to train pupils to "be honest, efficient, and self-reliant."[83] Upon the inauguration of the industrial building, new facilities were added including an ironworks department, a printing department, a photographic training school, and a very successful woodworking department led by Kim Paik Yun.[84] In 1920, the YMCA school was compartmentalized into six departments, including printing and bookbinding, soap- and candle-making, a "machine-shop," woodwork and cabinet making, rattan work, and photography.[85] The duration of the courses varied considerably. While it took pupils up to three years to master machine work or the making of cabinets and two years to learn printing or the making of soap and candles, those in photography were scheduled to graduate the course after six months.[86] After the first few months, a wage of six to nine yen could be paid to diligent and talented students, while for some courses, such as photography, participants had to pay a monthly fee of five yen.[87]

As a global organization, the YMCA could draw on financial resources stemming from donations by American businesses. Additionally, the Industrial School in Seoul seemed to have made some profit by selling products made by the students. Horace H. Underwood assessed the overall situation of the school in 1926 as being fairly good with well-equipped shops and "a competent staff of native instructors and workmen." Students and teachers attended chapel service every morning before work, and the school's graduates found employment at "bicycle and automobile repair shops, printing offices, photographers' studios, and other modern industries." At this juncture, Underwood also negatively pointed out the neglect of industrial training at other mission schools, which he deemed "a grave error in mission policy and a failure of missionary function."[88]

A few pages prior to the quote above, Underwood offered several criticisms, including that "academic education has been over-emphasized, the industrial, agricultural and technical sides being almost entirely untouched."[89] While colonial authorities generally encouraged industrial education by missionaries as long as they stuck to the regulations,[90] Underwood contended that this failure resulted from the mere necessity to support the pupils through self-help industrial departments.[91]

Contemporary observers clearly distinguished between the industrial school and the so-called self-help departments, which were purely meant to help pupils cover their educational expenses by working during their leisure time.[92] As Ritter stated in 1920: "Outside of the Japanese enterprises, the only industrial training in Korea is that offered by the missionaries, who are mostly American."[93] Most of these mission schools, however, did not specialize in industrial education, but rather maintained industrial education departments that provided basic industrial training while teaching the principles of industrial capitalism. Yet, in the 1910s and 1920s, numerous industrial education departments existed, which missionary accounts often commended as "an indispensable part of the institution."[94] Here, students were instructed in crafts such as sericulture, meat-curing and canning, or cloth-weaving.[95]

Apart from the YMCA school in Seoul, other such schools included the Mary Helm Industrial Department of the Holsten Institute and the Songdo Textile Department in Kaesŏng, both run by Southern Methodists, the Presbyterian Anna Davis School in P'yŏngyang, and the John D. Wells Academy for Christian Workers in Seoul, as well as the lace-making department attached to the girls' school at Kwangju, which was managed by the Southern Presbyterian Mission.[96] These institutions mainly taught their pupils how to manufacture textiles in factory-like settings, often using machines.[97] However, the Mary Helm Industrial Department and the lace-making department of the Kwangju girls' school emphasized handicraft, expressing a gendered conception of industrial education—in the eyes of Korean Christian leaders as well as Western missionaries, women were supposed to work at home and by hand.[98] On occasion, there seems to have been at least some degree of cooperation between the schools, such as that between the YMCA Industrial School and the Industrial School in Songdo.[99]

Another missionary school that operated an industrial department was the Hugh O'Neill Jr. Academy run by Presbyterians in Sŏnch'ŏn. Similar to the earlier example of the industrial department at Pai Chai College described by Isabella Bird Bishop, the school's self-help department was designed to help the students earn money for tuition and living expenses. In 1920, each department of the school was self-supporting.[100] Furthermore, the German Benedictine mission operated an industrial school at their monastery in Seoul, which later moved to Wŏnsan. About thirty Korean apprentices received two hours of theoretical classes and eight hours of practical training a day at the Sung-Kong-Handwerkerschule (Sung-Kong Craftsmanship School). Additionally, the weekly schedule comprised two hours of religious teaching, three hours of classical Chinese, three hours of mandatory Japanese language lessons, and four hours of math. During their last year of training, pupils furthermore attended bookkeeping class in order to enable them to run their own businesses after graduation.[101] Emphasis was put on carpentry and furniture-making, in particular,[102] but a blacksmith, locksmith, wheelwright, and tailor shop also existed at the monastery in 1913.[103] Training was strictly organized "systematically according to German standards," which reportedly entailed "great sacrifices" for the young Korean trainees.[104] This prejudiced view, contrasting German diligence with the supposed sloppiness of the Koreans, was echoed by American Protestant Underwood: "Emphasis [is] being laid on careful and workmanlike procedure, which is too likely to be neglected by the Oriental."[105] Looking back on a long monasterial tradition of craftsmanship, however, the motives of the European Roman Catholics for teaching industrial skills may have slightly differed from those of North American Protestants. Abbot Bonifatius Sauer summarized the intention behind the school as demonstrating that the Catholic Church was able to solve the social question. By creating a Catholic middle class that consisted of skilled craftsmen, he hoped to boost Catholic following in Korea.[106] Beyond teaching the skills of production and creating congregants, industrial education departments were instrumental in inculcating students with the principles of modern industrial capitalism based on the Protestant work ethic. The industrial education departments that were established and run by missionaries in this way not only proselytized Koreans but also aimed to thoroughly reform the bodies and minds of Korean converts.[107]

As exemplified by the German Benedictine Mission, Western missionaries in Korea were far from being a monolithic group. Nevertheless, the majority of missionaries, raised in nineteenth-century America, as well as Korean leaders of the YMCA, like Yun Ch'iho (1864–1945), believed that followers should maintain an industrious life and good conduct.[108] Moral virtues, including fortitude and self-control, had to be developed through hard work. Indeed, missionaries and Korean church leaders advocated a strong and productive work ethic because it "validated the sincerity of a person's faith and an individual's salvation," idealizing it as a sacred activity.[109] Moreover, just like the Korean proponents of self-strengthening, missionaries and Korean church leaders deemed industrialization a necessary requirement for attaining civilization and enlightenment.[110] In the 1920s and 1930s, the YMCA and the Presbyterian Church initiated a cooperative movement in rural Korea, influenced by international trends. They adopted the Danish cooperative model to promote self-sufficiency among laborers. Yet, contrary to the industrial education departments, the cooperative movement offered an alternative system of production, exchange, and consumption in the capitalist market economy.[111]

Conclusion

At the beginning of the twentieth century, Korea underwent a lengthy process of transformation, which also greatly affected the field of education. Prior to annexation, a number of Korean reformers recognized the need to embrace "modern" and industrial knowledge as a means of diminishing the influence of Confucianism on education and adapting to a globalizing world. Faced with the prospect of colonization by Japan, enhancing industrial competitiveness became crucial in order to safeguard national independence.

While some earlier Korean and missionary schools continued to exist with the tacit permission of colonial authorities after annexation, official Japanese education policy in Korea emphasized vocational and industrial education for Korean subjects to discourage them from pursuing higher education and to prevent them from becoming anti-Japanese. In the Korean context, the purpose of colonial education was twofold: to promote assimilation and maintain colonial rule, while also exploiting Koreans as inexpensive and acquiescent laborers to meet the economic demands of the empire. Consequently, official regulations favored industrial education over liberal arts education. Although missionaries occasionally shared a similarly patronizing perspective on Koreans, their primary motivation for providing industrial training lay in moral uplift, religious conversion, and societal advancement of their students. Throughout the colonial period, industrial education remained in demand even among Koreans supported by cultural nationalist educators advocating for effective national self-strengthening. This emphasis on industrial and technical education resurfaced after liberation in 1945, featuring prominently in the official educational agenda during Park Chunghee's so-called developmental dictatorship in the 1960s and '70s and its focus on human resource development.

Notes

1. Michael Kim, "The Colonial Public Sphere and the Discursive Mechanism of Mindo," in *Mass Dictatorship and Modernity,* ed. Michael Kim, Michael Schoenhals, and Yong Woo Kim (London: Palgrave Macmillan, 2013), 178–202.

2. For an overview, see E. Patricia Tsurumi, "Colonial Education in Korea and Taiwan," in *The Japanese Colonial Empire, 1895–1945*, ed. Ramon H. Myers and Mark R. Peattie (Princeton: Princeton University Press, 1984), 294.

3. See quote in Yoonmi Lee, *Modern Education, Textbooks, and the Image of the Nation: Politics of Modernization and Nationalism in Korea Education, 1880–1910* (London: Routledge, 2000), 8–9.

4. See for example Oliver Charbonneau, "Colonizing Workers: Labor, Race, and U.S. Military Governance in the Southern Philippines," *Modern American History* 4, no. 1 (2021): 25–47.

5. For the YMCA and the Social Gospel in general, see: Ian Tyrrell, "Vectors of Practicality: Social Gospel, the North American YMCA in Asia and the Global Context," in *Spreading Protestant Modernity: Global Perspectives on the Social Work of the YMCA and YWCA, 1889–1970*, ed. Harald Fischer-Tiné, Stefan Huebner, and Ian Tyrrell, Perspectives on the Global Past (Honolulu: University of Hawaii Press, 2020), 39–60; Timothy Yates, *Christian Mission in the Twentieth Century* (Cambridge: Cambridge University Press, 1994).

6. Albert L. Park, "A Sacred Economy of Value and Production: Capitalism and Protestantism in Early Modern Korea (1885–1919)," in *Encountering Modernity: Christianity in East Asia and Asian America*, ed. Albert L. Park, David K. Yoo, and Russell Leong (Honolulu: University of Hawai'i Press, 2014): 19–46.

7. One notable exception is Park, "A Sacred Economy of Value and Production."

8. For example, O Sŏngch'ŏl, *Singminji ch'odŭng kyoyuk ŭi hyŏngsŏng* [The formation of colonial primary education] (Seoul: Kyoyuk kwahaksa, 2000); on the relationship between missionaries and colonial authorities see Yoonmi Lee, "Religion, Modernity, and Politics: Colonial Education and the Australian Mission in Korea, 1910–1941," *Paedagogica Historica: International Journal of the History of Education* 52, no. 6 (2016): 596–613.

9. For the journal see Yoonmi Lee [Yi Yunmi], "Kyoyuksa yŏn'gu esŏ tamnonjŏng punsŏk ŭi ŭimi wa hwaryong: sŏn'gyosa charyo 'The Korea Mission Field' (1905–1941) ŭi tokpŏp ŭl chungsim ŭro" [Analyzing the historical text as a discourse: a reading of the 'The Korea Mission Field' (1905–1941)], *Han'guk kyoyuk sahak* 29, no. 1 (2007): 155–186.

10. Seungyop Shin, "A Thorn in Imperial Japan's Side: The Shifting Politics of Protestant Christianity in Colonial Korea," *Journal of Religious History* 46, no. 1 (March 2022): 130.

11. Tyrrell, "Vectors of Practicality," 39.

12. Park, "A Sacred Economy of Value and Production," 30.

13. Jean Comaroff and John Comaroff, *Of Revelation and Revolution: Christianity, Colonialism, and Consciousness in South Africa,* vol. I (Chicago: University of Chicago Press, 1991).

14. Rebekka Habermas, "Mission im 19. Jahrhundert—Globale Netze des Religiösen" [Mission during the 19th century—global networks of religion] *Historische Zeitschrift* 287, no. 3 (2008): 642–643.

15. Korean and Japanese names are rendered in the original order, family name first and given name last, throughout the article.

16. See Vipan Chandra, *Imperialism, Resistance, and Reform in Late Nineteenth-Century Korea: Enlightenment and the Independence Club* (Berkeley: University of California Press, 1988), 41.

17. *Isabella Bird Bishop, Korea and Her Neighbors: A Narrative of Travel, with an Account of the Recent Vicissitudes and Present Position of the Country* (New York: Fleming H. Revell Company, 1898), 387–388; also cited by Chandra, *Imperialism, Resistance, and Reform*, 61.

18. Albert L. Park, *Building a Heaven on Earth: Religion, Activism, and Protest in Japanese Occupied Korea* (Honolulu: University of Hawai'i Press, 2015), 39.

19. Chandra, *Imperialism, Resistance, and Reform.*

20. Klaus Dittrich, "The Beginnings of Modern Education in Korea, 1883–1910," *Pedagogic Historica: International Journal of the History of Education* 50, no. 3 (2014): 267.

21. Horace Horton Underwood, *Modern Education in Korea* (New York: International Press, 1926), 111.

22. Bishop, *Korea and Her Neighbors*, 210.

23. Underwood, *Modern Education in Korea*, 111.

24. Underwood, *Modern Education in Korea.*

25. Klaus Dittrich, "Korea's Internal Civilizing Mission: Education in the English Edition of *The Independent*, 1896–1898," *Acta Koreana* 16, no. 2 (2013): 431–472; see also Vipan Chandra, "Sentiment and Ideology in the Nationalism of the Independence Club (1896–1898)," *Korean Studies* 10 (1986): 23–24.

26. Henry Em, "Killer Fables: Yun Ch'iho, Bourgeois Enlightenment, and the Free Laborer," *Journal of Korean Studies* 25, no. 1 (2020): 50.

27. Frank M. Brockman, "Genesis of the Seoul Y.M.C.A.," *Korea Mission Field* 10, no. 2 (1914): 48–49.

28. Chŏn T'aekpu, *Han'guk kidokkyo ch'ŏngnyŏnhoe undongsa: Han'guk YMCA pansegi ŭi palchach'wi* [The history of the YMCA movement in Korea: Traces of half a century], Chŏn T'aekpu sŏnjip 4 [Collected works of Chun Taikpoo 4] (Seoul: Hongsŏngsa, 2017), 62–74.

29. See f.e. Kyoung Bae Min, *A Centennial History of the Seoul YMCA (1903–2003)* (Seoul: Seoul YMCA 2009): 65–67.

30. Yong-ha Shin, "The Sinminhoe's Independence Movement During the Last Years of the Chosŏn Dynasty," *Seoul Journal of Korean Studies* 7 (1994): 15.

31. For a more detailed account see: Andre Schmid, *Korea Between Empires, 1895–1919* (New York: Columbia University Press, 2002).

32. Chang Chiyŏn, "Manifesto of the Korean Association for Self-Strengthening," in *Sourcebook of Korean Civilization. Volume II. From the Seventeenth Century to the Modern Period*, ed. Peter H. Lee (New York: Columbia University Press, 1996), 415–416.

33. Shin, "The Sinminhoe's Independence Movement."

34. Pak Ŭnsik, "Essay on Education," in *Sourcebook of Korean Civilization. Volume II. From the Seventeenth Century to the Modern Period*, ed. Peter H. Lee (New York: Columbia University Press, 1996), 418–420.

35. Quoted from Pak Ŭnsik, "Essay on Technological Innovation," in *Sourcebook of Korean Civilization. Volume II. From the Seventeenth Century to the Modern Period*, ed. Peter H. Lee (New York: Columbia University Press, 1996), 420–422.

36. Shin, "The Sinminhoe's Independence Movement," 26–27.

37. See Chong-Sik Lee, *The Politics of Korean Nationalism* (Berkeley: University of California Press, 1963), 97.

38. The Government General of Chosen, *Manual of Education in Chosen* (Seoul: The Government-General of Chosen. Bureau of Education, 1920), 67.

39. Yi Man'yŏl, "Hanmal kumi cheguk ŭi Taehan sŏnkyo chongch'ek e taehan yŏn'gu" [A study of the Korean mission policies of the Euro-American nations during the late Chosŏn Dynasty], *Tongbang hakji* [Journal of the East] 84 (994): 44–45.

40. The Government-General of Chosen, *Annual Report on Reforms and Progress in Chosen (Korea) 1911–1912* (Seoul, 1912), 226, appendix B.

41. Takeshi Komagome, *Shokuminchi teikoku Nihon no bunka tōgō* [The cultural integration of the Japanese colonial empire] (Tōkyō: Iwanami Shoten, 1996), 92–96.

42. Kim, "The Colonial Public Sphere and the Discursive Mechanism of Mindo," 179.

43. Cited from Hiroshi Abe, "Higher Learning in Korea Under Japanese Rule: Keijo Imperial University and the 'People's University' Campaign," *Developing Economies* 9, no. 2 (1971): 175.

44. Ki-Baek Lee, *A New History of Korea*, trans. Edward W. Wagner (Cambridge: Harvard University Press, 1984).

45. The Government-General, *Annual Report on Reforms 1911–1912*, 226–228.

46. Theresa Bach, "Japanese Endeavor to Convert Koreans," *School Life* 5, no. 4 (1920): 14. This was clearly based on The Government-General of Chosen, *Manual of Education in Chosen*, 67–69.

47. The Government-General of Chosen, *The Third Annual Report on Reforms and Progress in Chosen (Korea) 1909–1910* (Seoul, 1910), 121.

48. The Government-General, *The Third Annual Report on Reforms 1909–1910*, 144.

49. The Government-General, *The Third Annual Report on Reforms 1909–1910*, 145.

50. Andrew Hall, "First Steps Towards Assimilation: Japanese-Run Education in Korea, 1905–1910," *Acta Koreana* 18, no. 2 (2015): 375.

51. Hall, "First Steps Towards Assimilation," 357.

52. Tsurumi, "Colonial Education in Korea and Taiwan," 295.

53. Johannes Witte, "Die Verschwörung in Korea und das Christentum" [The conspiracy in Korea and Christianity], *Zeitschrift für Missionskunde und Religionswissenschaft: Organ des Allgemeinen Evangelisch-Protestantischen Missionsvereins* [Journal of missiology and religious studies: Organ of the General Protestant Mission Association] 27 (1912): 181–183; and Frederick Arthur McKenzie, *Korea's Fight for Freedom* (New York: Fleming H. Revell Co., 1920), 220–221.

54. See McKenzie, *Korea's Fight for Freedom*, 220; in 1920, the Academy seemed to be up and running again with a focus on agriculture and sericulture; Richard H. Ritter, "Industrial Education in Korea," *North American Review* 212, no. 779 (October 1920): 529.

55. See Lee, *The Politics of Korean Nationalism*, 97.

56. Tsurumi, "Colonial Education in Korea and Taiwan," 306.

57. Abe, "Higher Learning in Korea Under Japanese Rule," 177.

58. Inaba Tsugio, *Chōsen shokuminchi kyōiku seisakushi no saikentō* [Reassessing the history of colonial education policy in Korea] (Fukuoka: Kyūshū Daigaku Shuppankai, 2010), 24.

59. The Government-General, *Manual of Education*, 64.

60. The Government-General, *Manual of Education*, 72.

61. See George Shannon McCune, "Honest Labor Makes the Man," *Korea Mission Field* 6, no. 9 (1911): 269; George A. Gregg, "Educating Hand and Brain," *Korea Mission Field* 8, no. 6 (1912): 170.

62. E. W. Koons, "The Schools of Seoul," *Korea Magazine* 2, no. 8 (1918): 39.

63. Leighanne Yuh, "Contradictions in Korean Colonial Education," *International Journal of Korean History* 15, no. 1 (2010): 315–317.

64. See Nitobe Inazō, "Japan as Colonizer," *Journal of Race Development* 2, no. 4 (1912): 361.

65. The Government-General, *Manual of Education*, 72.

66. Oh, Seong-chol [O Songch'ŏl] and Ki-seok Kim, "Expansion of Elementary Schooling Under Colonialism: Top Down or Bottom Up?" in *Colonial Rule and Social Change in Korea, 1910–1945*, ed. Hong Yung Lee, Yong-Chool Ha, and Clark W. Sorensen (Seattle: University of Washington Press, 2013), 114–139.

67. The Government-General, *Manual of Education*, 68.

68. Ritter, "Industrial Education in Korea."

69. For example, Yoshino Sakuzō, "Man-Kan o shisatsu shite" [Observations from Manchuria and Korea], in *Yoshino Sakuzō Chūgoku, Chōsen-ron* [Yoshino Sakuzō's treaties on China and Korea], ed. Matsuo Takayoshi, Tōyō bunko 161 (Tōkyō: Heibonsha, 1970), 83.

70. See Oh and Kim, "Expansion of Elementary Schooling Under Colonialism," 132.

71. O, *Singminji ch'odŭng kyoyuk ui hyŏngsŏng*, 309–310.

72. O, *Singminji ch'odŭng kyoyuk ui hyŏngsŏng*, 277–323.

73. Oh and Kim, "Expansion of Elementary Schooling Under Colonialism," 132.

74. The Government-General, *The Annual Report on Reforms and Progress in Chosen (Korea) 1938–1939*, 88–89.

75. Oh and Kim, "Expansion of Elementary Schooling Under Colonialism," 133.

76. Underwood, *Modern Education in Korea*, 118.

77. Ritter, "Industrial Education in Korea," 528.

78. Homer Bezaleel Hulbert, *The Passing of Korea* (New York: Doubleday, Page & Company, 1906), 341.

79. Chŏn, *Han'guk kidokkyo ch'ŏngnyŏnhoe undongsa*, 112.

80. The *Korea Mission Field* 7, no. 4 (April 1911), 96. Frank M. Brockman, "Young Men's Christian Association," *The Christian Movement*, 1915, 436.

81. See Yeon-seung Lee, "Between Nationalism and Internationalism: Yun Ch'i-ho and the YMCA in Colonial Korea" (PhD diss., Boston University, 2011), 407–408.

82. Koons, "The Schools of Seoul," 362.

83. Gregg, "Educating Hand and Brain," 170.

84. Brockman, "Young Men's Christian Association," 436–437.

85. Ritter, "Industrial Education in Korea," 528.

86. Koons, "The Schools of Seoul," 362.

87. Koons, "The Schools of Seoul"; Gregg, "Educating Hand and Brain," 173.

88. Underwood, *Modern Education in Korea*, 118; see also James Earnest Fisher, *Democracy and Mission Education in Korea* (New York: Columbia University, 1928), 91.

89. Underwood, *Modern Education in Korea*, 106.

90. ARRP 1935, 88

91. Underwood, *Modern Education in Korea*, 109.

92. Underwood, *Modern Education in Korea*, 111.

93. Ritter, "Industrial Education in Korea," 528.

94. McCune, "Honest Labor Makes the Man," 270.

95. Ritter, "Industrial Education in Korea," 529.

96. Underwood, *Modern Education in Korea*, 112–116.

97. See Park, *Building a Heaven on Earth*, 49.

98. Park, "A Sacred Economy of Value and Production," 35–36.

99. Gregg, "Educating Hand and Brain," 172.

100. Ritter, "Industrial Education in Korea," 529.

101. See You Jae Lee, *Koloniale Zivilgesellschaft: Alltag und Lebensweise der Christen in Korea (1894–1954)* [Colonial civil community: The everyday way of live of Christians in Korea (1894–1954)] (Frankfurt am Main: Campus Verlag, 2017), 145–149.

102. Underwood, *Modern Education in Korea*, 119.

103. See Lee, *Koloniale Zivilgesellschaft*, 146.

104. Quoted from Lee, *Koloniale Zivilgesellschaft*, 146.

105. Underwood, *Modern Education in Korea*, 119.

106. See Lee, *Koloniale Zivilgesellschaft*, 147.

107. Park, *Building a Heaven on Earth*.

108. Yun Ch'iho, "A Plea for Industrial Training," *The Korea Mission Field* 7, no. 7 (1911): 185–188.

109. Park, *Building a Heaven on Earth*, 48; Jang Wook Huh, "The Student's Hand: Industrial Education and Racialized Labor in Early Korean Protestantism," *Journal of Korean Studies* 25, no. 2 (2020): 364.

110. See Park, *Building a Heaven on Earth*, 48.

111. Quoted from Albert L. Park, "Reclaiming the Rural: Modern Danish Cooperative Living in Colonial Korea, 1925–37," *Journal of Korean Studies* 19, no. 1 (2014): 138.

PART III

Transformations and Legacies

CHAPTER 8

Pedagogies of International Development: Managing Black Nationalism and U.S. Racial Capitalism

Hossein Ayazi

Introduction

In his 1949 inaugural address, President Harry S. Truman outlined the Point Four Program, a plan to extend U.S. technical expertise for "the improvement and growth of underdeveloped areas."[1] Truman's announcement of the Point Four Program, and the 1950 establishment of the Technical Cooperation Administration within the Department of State to operate the program, characterized the response to the convergent crises facing the nation. Among these was the Soviet Union's emphasis on sustained racial disparities within the United States as part of its Cold War–era effort to undermine U.S. moral and political authority, and antiracist, anticolonial, and worker-led liberation movements across the Third World that continued to realize the prospect of a revolutionary-socialist internationalism.[2] These movements included the 1955 Bandung Conference, the 1957 Afro-Asian People's Solidarity Conference in Cairo, and the 1966 Tricontinental Conference of African, Asian, and Latin American Peoples in Havana. In this light, midcentury U.S. efforts to address national "underdevelopment" through bilateral agreements targeting agriculture, public health, and education, and through the racial integration of U.S. domestic institutions and foreign policy agencies alike, were self-serving. Such efforts, the architects of international development hoped, would help incorporate newly independent nations into U.S. circuits of accumulation while demonstrating to the world that the social relations of capitalist modernity were not hopelessly compromised by imperialist extraction, racial apartheid, and labor superexploitation.[3]

From the role of educators in the early twentieth-century work of "rural reconstruction" across Africa, Asia, and Latin America to the diffusion of technical expertise as a pillar of the Point Four Program itself, education has not simply been a set of discrete projects within the wider rubric of development work. Rather, education has been the *how* of development; how to, pedagogically, conduct a cooperative, cross-cultural process of engaging people in other societies (or subsets of U.S. society) in concrete rural development projects.[4] In this light, this chapter traces how emerging practices, institutions, and ideas of industrial education helped rationalize agrarian and rural space across the decolonizing Third World in ways that at once fostered the incorporation of newly independent nations into U.S. circuits of accumulation, and mystified the connections between imperialist extraction, racial apartheid, and labor superexploitation that characterized the United States' postwar ascendance. This chapter does so by interrogating how state, corporate, and philanthropic actors behind postwar industrial education constructed the "village" as a place of culturally attuned collaborations between local populations and foreign experts, and thus a site of counterrevolutionary developmental intervention. Focusing on the relationship between the United States and Liberia between the 1920s and 1960s, this chapter traces how industrial education was a vehicle through which such nested geographies of international development were constructed in opposition to ascendant Black (inter)nationalist and socialist Pan-Africanist movements within and beyond Africa.

Within Liberia, village-level industrial education as "development" was advanced by U.S. aid programs in partnership with private corporations and philanthropic entities in ways that subsumed and replaced missionary actors in the West African republic.[5] As the second Black republic in the world, after Haiti, such investments and partnerships around industrial education also functioned as a container for meditations on the meanings and practices of Black sovereignty and their relationship to the U.S.-led development paradigm. This was especially so by midcentury, as the prospect of national independence, political unity, and revolutionary-socialist struggle was being realized across Africa, and as the United States sought to contain such movements elsewhere in the world. This chapter argues that the terms of Black sovereignty for which Liberia was celebrated were recast into civic (i.e., non-racial or antiracist) ideologies based in the political and economic universality of the (U.S.-backed developed) nation, and that preeminent race-based industrial education institutions, programs, and partners were key to this process. Of particular significance were the Booker Washington Agricultural and Industrial Institute (BWI), founded in 1929 in Kakata, Liberia, and modeled after industrial education programs for Black youth in the U.S. South, and the U.S.-backed village-level work of teacher training, which Liberia supported by the 1950s and 1960s. It is this geopolitical post-race narrative and counterrevolutionary political horizon—developed through such industrial education efforts—that the Point Four Program had materialized internationally.

Recent scholarship has illuminated the geopolitics and geoeconomics of U.S.-led development and the role of experts in development practice.[6] Yet, undertheorized is the rise and consolidation of international development across the first half of the twentieth century in relation to the United States' early to mid-twentieth-century break with formal white supremacy and its adoption of an "official" antiracism that gave rise to processes of racialization that no longer neatly fit the "color line" that W. E. B. Du Bois had earlier theorized.[7] Amid the early and mid-twentieth-century proliferation of jointly antiracist, anticolonial, and worker-led movements, the need to mystify race as a means through which capitalist inequality is justified was clear, as was the hope that international development could do exactly that.[8] Indeed, the architects of the Point Four Program believed that it was necessary to bring about a lasting "Green Revolution" of agricultural productivity in the decolonizing world in order to stave off a "Red Revolution."[9] As historian and political theorist Charisse Burden-Stelly argues, this "Red Scare" was always already a "Black Scare,"[10] crystallized in the state-sanctioned panic against an ascendant Black (inter)nationalism and socialist Pan-Africanism. In this light, attention to how state, corporate, and philanthropic actors took the reins of industrial education in Liberia, in collaboration with Liberian authorities, affords insight into how the needs of twentieth-century U.S. state and private enterprise were met.

Recent scholarship on development has also illuminated the early to mid-twentieth-century construction of developmentalist economic geographies, from the "plantation economy" to the "national economy," as emergent instruments of racialization beyond the "color line."[11] Through these geographies of (under)development, formerly enslaved and colonized peoples were ascribed institutional and sociological "deficiencies" (unlike supposed biological deficiencies), which explained why they were "ill-prepared to make [the nation-state] work," as Adom Getachew argues, and thus why expert intervention was a condition of national independence.[12] Yet, the "village" has not been fully examined as another such geography of (under)development. Specifically, scholars who have studied its construction as an object of knowledge and governance have done so without attending to the counterrevolutionary shift in racial epistemology that characterized the United States' early to mid-twentieth-century ascendance.[13] What is the significance of village-level industrial education as "national development" during a period when state-recognized antiracisms would come to validate culturally powerful notions of the U.S. nation-state and its foremost interests? The type of village-level industrial education and "community development" modeled in Liberia, and the "international technical assistance" under the Point Four Program that materialized internationally, helped fashion a Black nationalism that was safe for empire.

The Booker Washington Agricultural and Industrial Institute: Developing a Nation of Industry and Skilled Workers

Celebrated as the second Black republic in the world, following Haiti, the West African Republic of Liberia was founded in the early nineteenth century by African American and Afro-Caribbean émigré-settlers, with support from the American Colonization Society. By the 1920s, Liberia's debt crisis put into question the republic's future and motivated official efforts to court foreign investment from private business and international and government agencies.[14] Emblematized by Liberia's 1926 granting to Firestone Natural Rubber Company a ninety-nine-year lease of up to one million acres of land for $0.05 per acre under actual development, foreign industry became instrumental to the success of the Black republic.[15] Harvey S. Firestone Jr. attributed the company's early success to what set the operation apart from prospective operations in colonial Philippines, colonial Sarawak, and even Mexico. Firestone claimed that the firm's success in Liberia was "largely dependent upon the organization of a permanent and contented labor force."[16] Harvey S. Firestone Jr. was certainly not alone in his understanding of such a labor force as a condition of the successful development of industry and nation alike. In October 1925, W. E. B. Du Bois wrote to Firestone's father, Harvey S. Firestone Sr., suggesting that in "this experiment in Liberia," not only might Firestone employ among his "industrial personnel . . . colored Americans of education and experience," but that the rubber operation could prove "that industry can do the same thing in a Black country like Liberia that it does in a white country like Australia: that is, invade it, reform it and uplift it by incorporating the native born into the imported industry and thus make the industry a part of the country."[17] Put simply, Liberia's success as a Black republic, and as a function of the success of the rubber industry in said republic, would hinge upon the incorporation of "the native born" into said industry.

When in 1930 a League of Nations commission of inquiry confirmed accounts that Firestone's recruitment practices had become akin to the enslavement and slave-trading of Liberia's "native born," it would be education—especially for youth—that would be affirmed as the way that U.S. industry could work for the Black republic, and vice versa. Lester Walton, a Black American journalist and commentator with *New York World*, and a soon-to-be-appointed U.S. Minister to the country, made the case in 1931 that a "wholesale withdrawal of Americans would close 108 of the 164 schools operating in the Republic. And without education, future generations of Liberians would be as helpless as their ancestors to create modern conditions."[18] The League of Nations agreed, recommending that progress around schooling continue because only then would "the educated native child not . . . [try] to escape from his debased environment and loaf about on the coast looking for a job in the Government offices, which are already overcrowded with Americo-Liberians."[19] At the heart of the promise of education was its capacity to be more culturally attuned to the needs of industry and the African wage laborer alike—the means of producing the industry-adjacent model villages that the League recommended, and eventually the means of carrying out village-level work directly. Key among such early education endeavors—the kind of coastal institution that was to bring migrant laborers in close harmony with what was touted as a national industry—was BWI, founded in 1929 in Kakata, Liberia. As Liberia's primary agricultural and industrial education project for the greater part of the first half of the twentieth century, BWI and the U.S. educators, reformers, and race relations experts that it brought together for village-level work would be instrumental in the making of a Black nationalism that worked for twentieth-century U.S. state and private enterprise.

BWI's story begins in 1922, when Americo-Liberian leadership considered the work of developing what Olivia Egleston Phelps Stokes of the Phelps Stokes Fund referred to as a "Tuskegee Institute in Liberia."[20] By the mid-1920s, Liberian President Charles D. B. King had visited the United States and toured the Tuskegee Institute. With the backing of the Phelps Stokes Fund, the Rockefeller Foundations General Education Board, the New York State Colonization Society, the American Colonization Society, and the Liberian government, an official charter for the "Booker Washington Agricultural and Industrial Institute of Liberia" was approved by the Liberian legislature in 1928, and the school officially opened in Kakata, Liberia, on March 17, 1929. Even before the League of Nations Commission, the case was being made for BWI to function as a Firestone-adjacent "model native village." As historian Donald Spivey argues, James L. Sibley, former State Agent of Negro Rural Schools in Alabama and the school's first president, had immediately sought to "bring BWI into complete harmony with the local tribes and villages and into a firm working relationship with the Firestone plantations."[21] Harvey Firestone Jr. himself echoed the commitment, vowing that "all efforts should be made to develop the Washington Institute."[22] His company provided BWI with specialists, buildings, machinery, and tools; entertained all visitors to BWI on the school's behalf; and funded BWI more than any other single entity.[23] BWI itself quickly became an engine of Liberian national development across the 1930s. In 1935, President Edwin Barclay announced plans to construct a network of roads—one from Monrovia through Kakata and the other from the Sierra Leone border to the Liberian coast—that would link Firestone's plantations and would be built by BWI students and non-BWI affiliated Indigenous (i.e., non-Americo-Liberian) peoples.[24] The rubber industry fueled the development of the nation and BWI alike. In 1936, BWI enrollment increased from sixty-eight students to over one hundred students, with no sign of slowing down.[25]

In the decades that followed the establishment of BWI and the Firestone plantation, it was the provision of technical training and "expertise" that shaped and legitimated U.S.-Liberian relations, even garnering the support of key critics and theorists of U.S. race and empire at the time. For example, W. E. B. Du Bois's famed 1933 response to the League's inquiry and findings was that "Liberia is not faultless . . . she lacks training, experience, and thrift"—what critic and theorist Cedric Robinson later noted as Du Bois's adherence to a "terrible species of technocracy."[26] How did this "terrible species of technocracy" under the leadership and in the interest of U.S. state and private enterprise subsume the racially hierarchical "civilizing mission" that industrial education had long been known for? U.S. philanthropies were key. Beginning in the 1920s and 1930s, U.S. philanthropies gave a platform to new ideas of race and race reform by heavily funding the social sciences—not only at white institutions such as the University of Chicago, the University of Pennsylvania, Yale University, and Columbia University, but also at Southern Black institutions, including Fisk University and Atlanta University.[27] As cultural theorist Jodi Melamed argues, between then and the 1960s, the philanthropy-backed social sciences promoted *cultural* rather than biological explanations of racial difference and inequity.[28] In doing so, the social sciences and their backers took racial inequality as an outcome of state and private enterprise and instead recast it as something to be *addressed* by state and private enterprise through intervention in the supposed cultural norms, deficiencies, and practices of racialized peoples.[29] Thus, through the "training, experience, and thrift" they afforded, what were originally envisioned as race-based industrial education institutions became key sites of the early to mid-twentieth-century rise of national and international development as exercises in an "official" antiracism—at once legitimating the expansion of U.S. state and private enterprise, and deflecting and limiting awareness of the logics of exploitation and domination in global capitalism.[30]

The U.S. South was an early forge of this "terrible species of technocracy." Once flagship institutions, the Hampton Institute and Tuskegee Institute abandoned their industrial training programs in the late 1920s because of dwindling enrollment.[31] The Rosenwald Fund—credited with building more than five thousand Tuskegee-inspired schools in the segregated South—aimed to help maintain the productivity of Southern plantation operations through advanced industrial training efforts, while avoiding any increased "competition among the races" that might arise from an increasingly skilled class of Black workers.[32] The solution posed by the Rosenwald Fund and others was simple: defining "skilled" work as practical knowledge to be supplemented by industrial education, such that "unskilled" work could be transformed by industrial education into "skilled" work, and workers could move from unskilled to skilled labor without changing occupations.[33] Critically, this act of redefinition as discourse and policy was rooted in the labor needs of local industry and aided by social scientists. According to education historian James D. Anderson: "First, the social scientists would conduct a study of the black population, its occupational structure, and projections of its industrial future in selected southern cities. Then the curriculum experts would use these findings to develop an industrial education program unique to the Negro job structure of each city."[34] Altogether, therefore, the proliferation of such schools for Black youth across the U.S. South at once exemplified their integration into U.S. democracy and deflected critique of the United

States' racially hierarchical political economy—a dynamic to be reproduced internationally, against the specter of an ascendant Black (inter)nationalism and socialist Pan-Africanism.

"An Educational System Truly Liberian": Making International Development Work for Empire

By the mid-twentieth century, industrial education had become a firmly established avenue of U.S.-led (inter)national development. Yet, for U.S. state and private enterprise and educators working within a Black republic like Liberia, there remained key challenges that BWI administrators believed the philanthropy-backed social sciences could help address. When the Rosenwald Fund was dissolved in 1948, Edwin Embree—a white American considered a national expert in race relations, the former vice-president of the Rockefeller Foundation, and the president of the Julius Rosenwald Fund beginning in 1926—helped establish the Liberian Educational Foundation, which planned for the "improvement of health, education, and living conditions" within Liberia.[35] At the same time, John Embree, anthropologist and son of Edwin Embree, was sent to Liberia to negotiate an agreement between the Liberian government and the United Nations for technical assistance.[36] Anticipating both Edwin and John Embree's work, Reverend Revington Embree, their cousin and principal of BWI beginning in 1941, asserted the need for social science disciplines for the school and nation alike. Embree's rationale was clear: Social scientific knowledge would facilitate improved exchanges with BWI students. As historian Donald Spivey argues, under Revington Embree's leadership, many students harbored a genuine dislike for BWI: "Liberian indigenes questioned the role of the school and its education program, which consisted mainly of chores and manual labor"—a genuine dislike that manifested in what Embree described as "problems with discipline."[37] Yet Embree interpreted such unrest as the result not of ongoing labor exploitation at the hands of BWI, Firestone, and the Republic, but of a gap in *cultural understanding.* As Spivey states, Embree believed BWI was faltering because its white U.S.-born administrators "could not understand [the teacher], the students, Liberia, and Africans in general."[38] As in the United States in the decades prior, philanthropy-backed social sciences would help meet the needs of U.S. state and private enterprise in Liberia by legitimizing this line of inquiry and the role of industrial education alike.

By midcentury, philanthropists and social scientists praised industrial education efforts like BWI because of the way they modeled effective cooperation with state and private enterprise alike in and across racially hierarchical contexts. After twenty-two years in the position, Rev. Dr. Anson Phelps Stokes announced that he would no longer be president of the Phelps Stokes Fund. Stokes used the occasion to reflect on the mission and accomplishments of the fund, publishing the 1948 report *Progress in Negro Status and Race Relations 1911–1946: The Thirty-Five Year Report of the Phelps-Stokes Fund*. According to the report, the Fund's mission was clear: assist interracial cooperation and improve the education of Negroes in America and in Liberia.[39] The Fund's successes were clear too: "the Phelps-Stokes Fund has . . . been able to secure a degree of cooperation from other agencies," including "the colonization societies in developing the Booker Washington Institute in Liberia [and] the State Department, through its Office of International Information and Cultural Affairs." Among such achievements, the report discusses the $15,000 provided by the Department of State for guidance and assistance by specialists in building, health, mechanics, and a community health program at BWI.[40] Another endeavor celebrating and encouraging such cooperation among state, corporate, and philanthropic actors in matters of education and race reform was the report *Africa Advancing: A Study of Rural Education and Agriculture in West Africa and the Belgian Congo*, published in 1945. The study was carried out by Thomas Campbell, the United States Department of Agriculture's field agent in charge of African American extension work in the seven lower Southern states; Jackson Davis, associate director of the General Education Board of the Rockefeller Foundation; and Margaret Wrong, who worked with the International Missionary Council.[41] Working across racially hierarchical contexts, the report studied industrial education in Belgian, French, and British territories alongside the Republic of Liberia, and praised the cooperation between educators and state, corporate, and philanthropic actors in the latter.

To the authors of these reports and across these contexts, the stakes of securing and emboldening educational and developmentalist endeavors like BWI were clear, as were the terms of struggle. As stated in the opening sentences of *Africa Advancing*: "Probably nothing in the present disturbed international situation has more potentialities for good or ill than the development of the so-called backward or dependent peoples of the world and their adjustment to western civilization. Of all the great areas in the world, Africa has remained the most isolated and least developed."[42] To Campbell, Davis, and

Wrong, ensuring that the advancement of Africa was safe and profitable for U.S. and other Western state and private enterprise first required framing Black (inter)nationalism and socialist Pan-Africanism as provincial and in fact a primary source of "ill." The authors state: "To advocate complete independence as a sure and speedy cure for all the ills of African life is a doctrinaire approach which feeds upon the fires of race prejudice."[43] Their second move was to valorize Western technical expertise and position its provision as an effective basis and necessary condition of national development, especially at the site of plantation production. The authors state that, within nations, "questions of distribution and of equitable sharing in the benefits of social and scientific advance have created tensions . . . Wide differences between sections of the population . . . *threaten the soundness of the national structure*. . . . These differences are emphasized in older agricultural areas where laborers and tenants follow the traditional plantation system."[44] Thus bringing familiar discourses of skilled labor to bear on the development of nation and industry alike in Africa, the authors state: "Science and the benefits of science must be applied in this field . . . Labor, if more highly skilled, can, by using technical knowledge, produce more and so raise the standard of living for all."[45]

During the 1950s and 1960s, Liberian President William V. S. Tubman's policy of "national unification" converged with this discourse on the national benefits of skilled labor—as both an anticommunist ideal and mystification of race as a procedure that justifies the non-generalizability of capitalist wealth. In 1954, at the First National Executive Unification Council in Harper, Liberia, President Tubman said: "We must now destroy all ideologies that tend to divide us. Americo-Liberianism must be forgotten and all of us must register a new era of justice, equality, fair dealing and equal opportunities for everyone from every part of the country regardless of tribe, clan, section, element, creed or economic status."[46] Centering industry and labor within this vision of national unity, Tubman soon established the Manpower Division of the Ministry of Planning and Economic Affairs to help identify and meet the Republic's labor needs as an exercise in "equal opportunities for everyone from every part of the country." In this way, the policy of "national unification" was one intended to construct an idealized Liberian civic nationalism as a moral ideal and disappear racialized dispossession and labor exploitation as conditions of Liberia's own national development. As in the United States, and as the authors of *Africa Advancing* envisioned for the continent, this Liberian national culture that Tubman celebrated was also a move against an ascendant Black (inter)nationalism and socialist Pan-Africanism and effort to contain the threat such movements might pose to Liberia's burgeoning national industries. By the end of the decade, under the banner of unification, Tubman stated this clearly: "We wage no war against socialism if it is kept within the territories and among people that are so inclined, but we shall fight till death any attempt to impose and force upon us what we consider a mystical illusion."[47]

The advancement of African nations would progress on such terms through an understanding of international provision of Western technical expertise through local schooling as an exercise in hybridization and cultural exchange rather than racialized imposition. "The real task in Africa," Campbell, Davis, and Wrong argue, "is to educate and develop the African, through self-discipline and responsibility, for self-government," and not disregard "the great advantage" that U.S. and European powers "offer for stability, continuity, and mutual aid in well-ordered plans of development towards the achievement of clearly defined aims."[48] With industry potential in mind, they argued that this work must begin with "agricultural research and education which aims at grafting modern scientific technique on to the primitive stock of native land custom."[49] Key U.S. proponents of Liberian "national unification" argued for the same while emphasizing the stakes of such work. As R. O'Hara Lanier, Project Director of the Phelps Stokes Fund and U.S. Minister to Liberia, stated in 1961, "national unification" could be achieved by "providing the transition from tribal Isolation to complete cultural and social integration of all segments of the population," and by wresting control of Indigenous education from "bush schools" without overriding Indigenous custom—addressing the supposed gap in *cultural understanding* that BWI president Revington Embree earlier lamented.[50] Doing so, Lanier argued, would stave off communist incursion. As he stated, the "problem of developing an educational system truly Liberian, in the African context . . . is a challenging one. It would be most desirable were the best of her many cultures incorporated into the educational system, while preserving the democratic Western outlook, for Liberians are known to be the most ardent devotees of the free enterprise system"—a problem that is "magnified since they are facing the expansion of the socialist and monolithic Marxist political economic systems in the newly independent countries."[51] Together, locally integrated industrial education was pushed as a key method of African national development against an ascendant Black (inter)nationalism and socialist Pan-Africanism.

Africans Teaching Africans: Village Schools and Counterrevolutionary Community Development

Imbued with anticommunist and ostensibly antiracist potential, a key method of U.S.-backed Liberian "national unification" and African national development in general was collaboration around industrial education that focused on training *teachers*. This meant a larger role for schools such as BWI. Indeed, as the authors of *Africa Advancing* state, "elementary schools in every village" and "secondary schools at strategic locations" would foster the application of "skill and intelligence to the ordinary activities of life."[52] Thus, the task at hand in Liberia was clear: "Liberia would have to look for an adequate supply of teachers for the small village schools"—"teachers qualified by training, expertise, and character"—and Liberia would have to "place Liberia College, the College of West Africa and Booker Washington Institute in a position to render the service that will be required."[53] This was the position held by the Liberian government too. As Nathaniel V. Massaquoi, Indigenous Liberian (Vai) and Secretary of Public Instruction from 1957 to 1962, stated in a 1960 article, "The Unification Question and Education in Liberia," "The impact of industrialization and urbanization upon the tribal people can be considered as a transition in the evolution of the people from an underdeveloped, primitive tribal to an industrial and diversified society; only a diversified educational system with its foundation thoroughly wedded to the life of the villages and smaller communities will be able to absorb this shock, this transition."[54] It is for these reasons, and the reasons put forth by Liberia's U.S. backers, that Massaquoi regarded the government's role in securing competent teachers in sufficient numbers for the nation's expanding school system as the Republic's "number one problem in education."[55]

An international endeavor, twenty percent of the Republic's revenue between 1952 and 1961 was directed toward joint efforts at education and welfare with the Joint Commission of the United States and Liberia. Although BWI benefited from such investments and expanded accordingly (adding eleven new courses that decade), the growing focus on teacher training hastened the rise of other educational institutions as the premier institutions of international collaboration and Liberian national development. In 1947, the Liberian government established the first degree-granting teacher training college within the College of West Africa (CWA). In 1950, the program was separated from CWA, and in 1951 it became the W.V.S. Tubman's Teachers College of the University of Liberia—its main function being the training of elementary and secondary school teachers. Soon after, in 1952, the U.S.-based board of BWI turned the institution over to the government-run university. By 1954, BWI became the Booker Washington Agricultural and Industrial College of the University of Liberia, and operated as the industrial arm of the school—the first and only national Liberian government secondary school in existence at the time, yet no longer the basis of the role played by U.S. state and private enterprise in Liberian national development.[56] For example, BWI's U.S. sponsors and the Firestone Natural Rubber Company itself began a phased decrease of financial and administrative support. Firestone even initiated its own training program for workers rather than rely on BWI.[57]

This shift toward state-run and locally operated teacher training institutions in Liberia corresponded with the emergent form and function of U.S. international technical assistance. U.S. President Harry S. Truman's 1949 announcement of the Point Four Program ushered in a "cooperative" program of technical assistance and self-help administered under the Technical Cooperation Administration (TCA).[58] In 1949, Dr. Henry Garland Bennett, head of the Point Four Program, stated that it was to be "primarily a program of teaching and demonstration" that would deal "directly with villages and with people" and would help people to "develop simple improvements in methods and practices which they themselves will understand and carry forward." This would be "progress through self-help" that "adapted old methods of demonstration and training to local situations" and put in people's hands "the kinds of tools that they can use now to help themselves" achieve "practical results" for their immediate "economic and social lives."[59] This commitment to the work of grassroots-level education was considered an exercise in nation-building, connecting national governments and local trainees to villagers in a bottom-up process of collective problem-solving that increased popular support for government officials who would in turn feel "responsibility for the well-being and welfare of the people."[60] That such work would also be considered instrumental in the fight against communism should come as no surprise. Echoing the stakes of national unification that Liberian officials made clear, Bennett asked in that same speech, "Who will teach these people—we, or the communists?"[61]

During the 1950s, and across key administrative changes, the Point Four Program would retain its grassroots-level approach to national development and international collaboration. By the fall of 1953, the TCA was subsumed into the Foreign Operations Administration (FOA), which in turn created a Community Development division in 1954. By 1956, the FOA's replacement, the International Cooperation Administration (ICA), had aid and staff for forty-seven community development programs in twenty-three countries. As the administrative apparatus of the Point Four Program was sharpened, so too was the "village" as the administrative unit of such work within partner countries.[62] Rather than sending multiple technicians from multiple administrative units, advocates of "community development" under the Point Four Program exalted multipurpose "grass roots village workers."[63] As Carl C. Taylor, a rural sociologist and FOA/ICA community developer, stated in 1956, the local expert was to be a person "born and reared in local villages."[64] Even further, the village is where they would carry out such work, for they were to be "the extension agent at the village level for all the government services and the energizer of village self-help" across the spectrum of village life. Through state-managed and locally integrated teacher training, and through the policy of "national unification," Liberia was perfect for cultivating the practice of U.S.-led international development in service of U.S. state and private enterprise, and against an ascendant Black (inter)nationalism and socialist Pan-Africanism. This model relationship between the United States and Liberia soon took shape through Liberia's Rural Teaching Training Institutes (RTTI).

By 1956, the need for dedicated rural teacher training institutions was being discussed in earnest in Liberia.[65] That year, the Liberian legislature passed a bill creating training centers for elementary school teachers, with the hope of attaching these to regular high schools so that secondary academic subjects as well as teacher training could be taught. The plan came to fruition with the 1958 agreement to establish at least one RTTI that would offer a secondary education alongside training for rural primary teachers.[66] U.S. backing soon followed, as a joint U.S.-Liberian commission advised that an institution in the United States be contracted to provide necessary training personnel. In 1959, the Secretary of Public Instruction and the Assistant Director of the United States Overseas Mission visited several institutions in the United States to select one to help in the development of the RTTI. After a survey, Tuskegee Institute in Alabama was selected for this purpose.[67] In 1960, an agreement was reached between the Liberian and U.S. governments to contract with Tuskegee Institute to staff and train rural teachers at the RTTI in Zorzor, and possibly a second RTTI at Kakata, where BWI was located.[68] Critically, the agreement between Liberia and the United States, and the partner institutions across both countries, was mediated by the ICA. Further, the ICA was to finance the construction of the schools, provide technical assistance for training Liberian counterparts and for developing educational materials, and supply permanent equipment and items such as the power and water plants.[69]

The international teacher training and community development endeavor quickly bore fruit. In 1963, Zorzor Rural Teaching Training Institute graduated its first class of fifty-two students, and in 1964 Kakata RTTI was opened.[70] Yet the success of the endeavor extended beyond the number of teachers trained. The U.S.-backed RTTIs within Liberia were the first such exercise of bottom-up counterinsurgency in Africa—a pillar of the U.S. postwar activity across other colonial and postcolonial contexts.[71] This counterrevolutionary construction of teacher training and community development remained a pillar of U.S. Cold War–era policy. In Vietnam, for example, government pacification efforts reprised community development's combination of security and participatory development.[72] Critically, the United States was not alone in its commitment to counterrevolutionary village-level national development. For example, the British were using community development techniques as pacification in Malaya, Kenya, and elsewhere—efforts that inspired Ngo Dien Diem in South Vietnam, and that the United States soon supported.[73] Yet, what remained unique about the U.S. role in Africa—by way of Liberia in particular—was that such work required a very real and very persistent reckoning with the terms of Black (inter)nationalism and revolutionary-socialist struggle through which African independence was fought, and against which U.S. nationalism and international technical assistance were constructed.

Conclusion: African Unity, Liberian Style

As anticolonial nationalist ferment grew in Africa during the mid-twentieth century, the debates over the future form of political organization of the African continent were imbued with questions of how this form of political organization might rebuff or reconstitute historical relations of colonial dependency within an emergent postcolonial era. Between 1958 and 1962, the debate over how to forge a continent-wide international organization unfolded across a number of conferences in Ghana, Liberia, and Nigeria.[74] Across these meetings, two informal associations of African states formed: the Monrovia Group (officially the "Conference of Independent African States") and the Casablanca Group. Representing the interests of the latter group, Kwame Nkrumah, Prime Minister of Ghana and noted political theorist and revolutionary-socialist figure, argued for a tight political federation of states. His reasoning was that total political and economic freedom from European domination could occur only through the formation of a strong unified African political entity that could negotiate with major powers and defend itself militarily if needed. Representing the interests of the former group, Liberia's President William V. S. Tubman argued for a much looser association of independent sovereign states with strong economic ties, but without the need to submit to another political authority. Ultimately, these two groups converged in Addis Ababa, Ethiopia, in 1963, which became the founding conference of the Organization of African Unity. At the meeting, thirty-three African governments signed the charter whose contents reflected Tubman's argument for an intergovernmental international organization instead of Nkrumah's socialist vision of a political union.

At the same time, in a 1960 meeting between President Tubman and Deputy Under Secretary for Administration Loy W. Henderson—that took place at the request of U.S. President Dwight D. Eisenhower and U.S. Secretary of State John F. Dulles, and that addressed the topic of U.S. representation in West Africa amid the discussions of African political unity—Tubman asserted that the promise of his vision of internationalism could be evidenced by U.S. support of Liberian national development if such support was fully extended. Henderson recounted Tubman's frustration in a telegram: "[Tubman's] general theme was that although . . . Liberia had looked to US as its mentor and protector; had in general followed US lead in international affairs; had entrusted to US private enterprise development of Liberian resources nevertheless US had not given Liberia assistance [of the] kind needed enable it keep up with its neighbors which although colonies were being developed culturally, economically, and militarily by their mother countries."[75] Henderson understood the dilemma with regard to the possible trajectories of African political unity: "As [the] rising tide of African nationalism makes [an] impact on Liberia, [Tubman's] position will become progressively weakened unless he can demonstrate it pays [to] be ally of US." Henderson thus agreed with the preconditions of national progress and the need to put such progress on display for other African nations: "[National] progress [is] not possible without vigorous U.S. aid," a point that would need to be made with "extreme care."[76]

In light of such negotiations within Africa, and between African nations and the United States, what helped produce the victory of Tubman's vision over Nkrumah's? Using Liberian national development as a model, Tubman reasoned that continued cultural, economic, and political affiliation with former "mother countries" was unproblematic.[77] The persuasiveness of Tubman's argument followed from the construction of the developed nation as that which could supposedly foster social forms of security through the market, and do so against racial animosities of the past—a political horizon that the discourse of Liberian "national unification" framed, and that U.S.-led "community development" and "international technical assistance" could materialize internationally.[78] The educational endeavors at the heart of U.S.-Liberian collaborations, and that Tubman would reference, hinged upon the village as a locus of training and supposed cross-cultural exchange, and did so through and against an ascendant Black (inter)nationalism and socialist Pan-Africanism that inflected ongoing struggles for independence and political unity within and beyond Africa. In this way, and as this chapter has argued, U.S.-backed industrial education efforts in Liberia between the early and mid-twentieth century gave rise to official discourses of national development and international technical cooperation that disappeared race as a referent for the historical development of modern capitalism and the uneven nature of Liberia's own national development. This counterrevolutionary rearticulation of racial capitalism, nation, and empire was the epistemological foundation upon which political affiliation with former "mother countries" could be deemed unproblematic.

Notes

1. "Inaugural Address of Harry S. Truman," January 20, 1949, Yale Law School, The Avalon Project, https://avalon.law.yale.edu/20th_century/truman.asp.

2. Jodi Melamed, *Represent and Destroy: Rationalizing Violence in the New Racial Capitalism* (Minneapolis: University of Minnesota Press, 2011), 5; Adom Getachew, *Worldmaking After Empire: The Rise and Fall of Self-Determination* (Princeton: Princeton University Press, 2019), 143.

3. David Nally and Stephen Taylor, "The Politics of Self-Help: The Rockefeller Foundation, Philanthropy and the 'Long' Green Revolution," *Political Geography* 49 (2015): 52; Alyosha Goldstein, "On the Internal Border: Colonial Difference, the Cold War, and the Locations of 'Underdevelopment,'" *Comparative Studies in Society and History* 50, no. 1 (2008): 26–56.

4. Francis Bonenfant-Juwong, "Teacher-Technicians: Progressive Education, Point Four, and Development," *Diplomatic History* 45, no. 5 (2021): 985.

5. As Bronwen Everill argues in this volume, U.S. missionary work in Liberia was instrumental to the industrial school movement's global takeoff in the late nineteenth and early twentieth centuries.

6. Deborah Cowen and Neil Smith, "After Geopolitics? From the Geopolitical Social to Geoeconomics," *Antipode* 41, no. 1 (2009): 25; Nicole Sackley, "The Village as Cold War Site:

Experts, Development, and the History of Rural Reconstruction," *Journal of Global History* 6, no. 3 (2011): 483.

7. Such processes of racialization, according to cultural theorist Jodi Melamed, involve "convert[ing] the effects of differential value-making systems into normative orders of difference" that make it possible to order, analyze, describe, and evaluate what emerges out of force relations as the permissible content of other domains of U.S. modernity (e.g., law, politics, economy). Melamed, *Represent and Destroy*, 41; Nikhil Pal Singh, *Black Is a Country: Race and the Unfinished Struggle for Democracy* (Cambridge: Harvard University Press, 2005), 143 and 150.

8. Marking the beginning of such work, the League of Nations helped excise the anticapitalist implications of anticolonial struggle in the wake of the Russian Revolution, and the League's mandates systems recast self-determination itself as a racially differentiated principle compatible with existing relations of economic dependence. Getachew, *Worldmaking After Empire*, 39; Melamed, *Represent and Destroy*, 15.

9. Nick Cullather, *The Hungry World: America's Cold War Battle Against Poverty in Asia* (Cambridge: Harvard University Press, 2010), 179.

10. Charisse Burden-Stelly, *Black Scare/Red Scare: Theorizing Capitalist Racism in the United States* (Chicago: University of Chicago Press, 2023).

11. See: Hossein Ayazi, "Land Reform, Race Reform: Interwar Anticommunism and U.S. Racial Capitalism," *Environment and Planning D: Society and Space* 40, no. 5 (October 2022): 900–922; Timothy Mitchell, "Economists and the Economy in the Twentieth Century," in *The Politics of Method in the Human Sciences: Positivism and Its Epistemological Others*, ed. George Steinmetz (Durham: Duke University Press Books, 2005), 126–141; Daniel Speich, "The Use of Global Abstractions: National Income Accounting in the Period of Imperial Decline," *Journal of Global History* 6, no. 1 (2011): 7–28.

12. Getachew, *Worldmaking After Empire*, 26.

13. Sackley, "The Village as Cold War Site," 483.

14. Whyte, "A State of Underdevelopment," 30.

15. Frank Robert Chalk, "The United States and the International Struggle for Rubber, 1914–1941" (PhD diss., University of Wisconsin, 1970), 152.

16. Wayne Chatfield Taylor, *The Firestone Operations in Liberia* (Washington, D.C.: National Planning Association, 1956), 66.

17. W. E. B. Du Bois, *The Correspondence of W.E.B. Du Bois, Volume I. Selections, 1877–1934*, ed. Herbert Aptheker (Amherst: University of Massachusetts Press, 1997), 322–323; cited in: Cedric Robinson, "DuBois and Black Sovereignty: The Case of Liberia," *Race & Class* 32, no. 2 (1990): 46.

18. Lester A. Walton, "Liberia Suffers Bad Sanitary Conditions," *New York World*, February 1, 1931; cited in: Susan Curtis, *Colored Memories: A Biographer's Quest for the Elusive Lester A. Walton* (Columbia: University of Missouri Press, 2008), 212.

19. "Report of the Committee of the Council of the League of Nations," 17.

20. Although the brainchild of Olivia Egleston Phelps Stokes, it was an idea that was already floated by Anson Phelps Stokes and Booker T. Washington himself as early as 1910. Donald Spivey, *The Politics of Miseducation: The Booker Washington Institute of Liberia, 1929–1984* (Lexington: University Press of Kentucky, 1986), 19.

21. Spivey, *The Politics of Miseducation*, 125.

22. Spivey, *The Politics of Miseducation*, 127.

23. Spivey, *The Politics of Miseducation*, 40.

24. Spivey, *The Politics of Miseducation*, 42.

25. BWI came to represent most of the republic's Indigenous peoples—namely, Bassa, Buzi, Kpelle, Gbande, Kru, Mano, Mandingo, Grebo, Gizi, Gola, Akra, Guio, Kwana, Gibi, and Kongo peoples.

26. Robinson, "DuBois and Black Sovereignty," 44.

27. Melamed, *Represent and Destroy*, 19.

28. Melamed, *Represent and Destroy*, 19.

29. Racial liberalism also rendered Black socialist internationalism a reflection of Black cultural pathology, itself constructed as maladjustment to racism. Melamed, *Represent and Destroy*, 60–61.

30. Melamed, *Represent and Destroy*, 57.

31. James D. Anderson, *The Education of Blacks in the South, 1860–1935* (Chapel Hill: University of North Carolina Press, 1988), 234.

32. Anderson, *The Education of Blacks*, 215–216.

33. Anderson, *The Education of Blacks*, 206, 215–216.

34. Anderson, *The Education of Blacks*, 208.

35. Alfred Perkins, *Edwin Rogers Embree: The Julius Rosenwald Fund, Foundation Philanthropy, and American Race Relations* (Bloomington: Indiana University Press, 2011).

36. Alfred Métraux, "UNESCO and Anthropology," *American Anthropologist* 53, no. 2 (1951): 294–300.

37. Spivey, *The Politics of Miseducation*, 65.

38. Spivey, *The Politics of Miseducation*, 66.

39. George J. Hill, "Intimate Relationships: Secret Affairs of Church and State in the United States and Liberia, 1925–1947," *Diplomatic History* 31, no. 3 (2007): 500.

40. Hill, "Intimate Relationships," 500.

41. Mona Domosh, "Race, Biopolitics and Liberal Development from the Jim Crow South to Postwar Africa," *Transactions of the Institute of British Geographers* 43, no. 2 (2018): 312.

42. Jackson Davis, Thomas M. Campbell, and Margaret Wrong, *Africa Advancing: A Study of Rural Education and Agriculture in West Africa and the Belgian Congo* (New York: Friendship Press, 1945), 7.

43. Davis, Campbell, and Wrong, *Africa Advancing*, 83.

44. Davis, Campbell, and Wrong, *Africa Advancing*, 86.

45. Davis, Campbell, and Wrong, *Africa Advancing*, 86.

46. "President Tubman's Address at the First Executive Council," 1954, William V. S. Tubman Papers, 1904–1992.

47. "President Tubman's Opening Address Delivered to the Third Biennial Unification Council," 1963, William V. S. Tubman Papers, 1904–1992.

48. Davis, Campbell, and Wrong, *Africa Advancing*, 88.

49. Davis, Campbell, and Wrong, *Africa Advancing*, 88, 97.

50. R. O'Hara Lanier, "The Problem of Mass Education in Liberia," *Journal of Negro Education* 30, no. 3 (1961): 256.

51. Lanier, "The Problem of Mass Education in Liberia," 260.

52. Davis, Campbell, and Wrong, *Africa Advancing*, 28–29.

53. Davis, Campbell, and Wrong, *Africa Advancing*, 28–29.

54. Massaquoi is clear which industries these are: "the rubber plantations, Liberian or Firestone owned, and the mines of the LMC and LAMCO." Nathaniel V. Massaquoi, "The Unification Question and Education in Liberia," *Phi Delta Kappan* 41, no. 4 (1960): 151–152.

55. Nathaniel V. Massaquoi, "The Fifth Annual Report of the Department of Public Instruction for the Period October 1, 1959 to September 30, 1960" (Monrovia: Department of Public Instruction, 1960).

56. Harry Fumba Moniba, "Booker T. Washington, Tuskegee Institute, and Liberia: Institutional and Moral Assistance, 1908–1969" (PhD diss., University of Rochester, 1975), 124.

57. Spivey, *The Politics of Miseducation*, 145.

58. Bonenfant-Juwong, "Teacher-Technicians," 992.

59. "Memorandum by the Technical Cooperation Administrator (Bennett) to the Director of the Management Staff (Heneman)," April 20, 1951, A/MS Files, Lot 54 D 291, TCA Program, U.S. Department of State, Office of the Historian; cited in: Bonenfant-Juwong, "Teacher-Technicians," 997.

60. Bonenfant-Juwong, "Teacher-Technicians," 998.

61. Henry Garland Bennett, "Some Notes on Point 4," 1952, NX5469; PAM HC59, Technical Cooperation Administration, U.S. Department of State; Cited in: Bonenfant-Juwong, "Teacher-Technicians," 997.

62. Sackley, "The Village as Cold War Site."

63. Bonenfant-Juwong, "Teacher-Technicians," 1003.

64. Carl C. Taylor, "Community Development Programs and Methods," *Community Development Review*, December 1956, 40; cited in: Bonenfant-Juwong, "Teacher-Technicians," 1003.

65. Yet, the Fundamental Educational Center at Klay was established in 1953 and jointly operated by the Liberian government and UNESCO.

66. KRTTI would serve the Central Province. ZRTTI would serve the Western Province, and a third RTTI in Webbo was to serve the Eastern Province, though WRTTI never opened. Charles Arthur Mehaffey, "Teacher Education in Liberia" (PhD diss., Bowling Green State University, 1980), 133.

67. Mehaffey, "Teacher Education in Liberia," 137.

68. As national government schools, these institutions reported to the Ministry of Education rather than the county supervisor of schools; Mehaffey, "Teacher Education in Liberia," 137.

69. Mehaffey, "Teacher Education in Liberia," 138.

70. "Annual Report of Secretary of Education, R. L. to the Honourable the Legislature of the Republic of Liberia in the First Session of the Forty-Fifth Legislature of the Republic of Liberia, October 1, 1962 to September 30, 1963" (Liberia Department of Education, 1963), 40; cited in: Mehaffey, "Teacher Education in Liberia," 137.

71. For example, in the Philippines in the early 1950s, U.S. and Philippine authorities "weaponized" community development as counterinsurgency during the communist Huk rebellion. See: Chapter 4 in Daniel Immerwahr, *Thinking Small: The United States and the Lure of Community Development* (Cambridge: Harvard University Press, 2015).

72. Bonenfant-Juwong, "Teacher-Technicians," 1009.

73. Sackley, "The Village as Cold War Site," 498; cited in: Bonenfant-Juwong, "Teacher-Technicians," 1009.

74. These were the All-African People's Conference of 1958 in Ghana, the Sanniquellie Conference of 1959 in Liberia, the Monrovia Conference of 1959 in Liberia, and the Lagos Conference of 1962 in Nigeria.

75. Loy W. Henderson, "Telegram from the Embassy in Senegal to the Department of State Dakar," 1960, U.S. Department of State, Office of the Historian.

76. Henderson, "Telegram from the Embassy in Senegal."

77. Chandra R. Dunn, "Africa and Liberia in World Politics: An Analysis of Liberian Foreign Policy During the 20th Century" (PhD diss., American University, 2016), 203.

78. For example, in a report on the 1958 All African People's Conference, U.S. civil rights activist George M. Houser noted that "Whereas until recently Pan-Africanism has always been a racial concept ('Africa for the Africans'—meaning the black Africans), now a residential element has been added. Anyone living in Africa, white or black, could be a part of the Africa of the future so long as the basic principle of democracy ('One man one vote') is accepted." George M. Houser, "A Report on the All African People's Conference Held in Accra, Ghana" (New York: American Committee on Africa, 1958), 2–3.

CHAPTER 9

In the Name of Rationality and Progress: Boarding School Education in the Soviet Arctic

Lukas Allemann

Introduction

Boarding schools have been the preferred form of schooling for Indigenous minorities in all circumpolar nation-states (Canada, the United States, Finland, Norway, Sweden, and the Soviet Union/Russia).[1] In this chapter I show that "industrial education" for the Indigenous minorities of the Soviet Arctic meant, not surprisingly, something different than in capitalist societies—and yet with striking similarities at a fundamental level. While industrial education as a term was absent in the Soviet Union, the domination of rural ethnic minority populations through educational and large-scale economic experiments nevertheless shows the underlying closeness of Soviet educational policies and "Western" approaches that openly pursued industrial education for marginalized social groups. Contrary to industrial education in noncommunist countries, in the Soviet Union the curriculum did not officially differ depending on social class or race and had the goal of providing equal educational and professional opportunities to everyone. However, as I show in the following case study, for many ethnic minorities the educational opportunities on offer did in practice perpetuate inequality, and came at the cost of losing Indigenous livelihoods, knowledge, languages, and self-esteem. I also argue that the Soviet Union should be seen as a thoroughly "Western" state when it comes to a core ideological conviction: its belief in rationalization and technical progress as the means to subdue resourceful places and "uncivilized" people. Thus, in their education policies, all circumpolar states—no matter whether capitalist or communist—were led by the thought of bringing "rationality" and "progress" to their imperial subjects. In the Soviet Union, this was reached *not* by designing a special curriculum for Indigenous children, but by applying a single standardized curriculum for the first eight years of schooling across the entire country, disregarding Indigenous languages and livelihoods.

Empirically, this chapter focuses on the Soviet policies toward the Indigenous people of the Kola Peninsula (also known as Russian Lapland or Murmansk Region, Northwest Russia), the Sámi, who were strongly affected by different forms of state-induced displacement. The Sámi people are Indigenous people inhabiting the region of Sápmi, which encompasses the northern parts of contemporary Norway, Sweden, and Finland, as well as the most northwestern tip of Russia. Estimations of their number oscillate between 60,000 and 100,000. The smallest population of the Sámi, around 1,600 people, lives in Russia.

The insights from this chapter will allow us to see the global commonalities of bringing formal education to minorities as a tool to dominate. I will try to deepen the idea of the Soviet Union as a "Western" power in the sense of its high-modernist affinity to technical and scientific progress and deeply European ideological roots, as earlier suggested by the anthropologist James Scott.[2] I set out to breathe more empirical life into this idea by relying on oral history research among the Sámi people in Russia, which demonstrates that such affinity was very visible in the field of education.[3]

The intellectual roots of Marxian thought are located in Europe. Nevertheless, in connection with the Iron Curtain divide resulting from the Cold War, a steady othering of the Soviet Union took place in the press and scholarship of the capitalist bloc throughout the twentieth century, which framed the Soviet Union as the complete antipode of anything "Western." I will adhere here to Tim Ingold's sharp-minded thoughts on the contested terms "Western" and "modern." He writes a convincing "apology" for using both terms in spite of all the generalizations and impreciseness they entail, where possible legitimate contestations of "Western" and "modern" are the following: first, in history, anthropology and other scholarship have often served as a generalizing foil against an equally essentialized "native point of view," or "Soviet point of view." Second, much of the philosophical critique of so-called Western or modern thought stems itself from Western lines of thought. In other words, Western traditions of thought are very heterogeneous and cannot be exhaustively subsumed under a generalizing "Western" foil.[4] I add that, third, Marx and Engels's writings, as the ideological foundations and economic rationale upon which the Soviet experiment rested, represent one of the most prominent Western schools of thought.

This fact is often obscured by the political differences that emerged between the "East" and the "West" in the course of the twentieth century. As Ingold notes, "our very [scholarly] activity, in thinking and writing, is underpinned by a belief in the absolute worth of disciplined, rational inquiry. . . . It is to this belief that the terms 'Western' and 'modern' refer."[5] Understanding this belief is relevant for seeing industrial schooling and other forms of social engineering as a common characteristic of what Scott called high-modernist societies, across the Western-Soviet or capitalist-communist divide. Indeed, it was exactly on account of the Soviet Union's belief in discipline and rationality that it was thoroughly "Western." A discussion of the Soviet state's schooling policies toward its Indigenous minorities—in this chapter based on the Russian Sámi as a case study—enables us to lay bare some ideological commonalities between "Western" nations and the Soviet state hidden behind the more prominent ideological veil of difference caused over many decades by capitalism-communism dichotomies. The chapter focuses on these macro-structural parallels, arguing that, in this respect, the Soviet Union functioned in similar ways as a "Western" nation.

Historical Context

The Sámi were in contact with Russian culture earlier than the Indigenous people of Siberia, with trade relations established since roughly the thirteenth century and Christianization starting from the fifteenth century. This makes it hard to speak of a sudden colonization.[6] In the nineteenth century, Komi and Nenets people, two ethnic groups originating from territories closer to the Urals, migrated to the Kola peninsula in search of new pastures. Despite a history of competing with the Sámi for pastures and other resources, during the Soviet period, intermarriage and co-working in collective farms became common.[7] However, before Soviet state authority in the Arctic was firmly established in the 1930s, the Sámi kept their semi-nomadic lifestyle governed by the yearly migration cycle of reindeer. Only the Soviet authorities put an end to this lifestyle. The state was concerned with fixing the supposed "problems" it identified as inherent to Indigenous peoples living in this area. The "solutions" included sedentarizing people, collectivizing their economic activities, and turning a subsistence economy based on multiple livelihoods into a large-scale monoculture—in short, reshaping the rural Indigenous populations according to high-modernist understandings of governance. Sending Indigenous children to boarding schools was a central component of these broader policies.

In connection with these developments, the state organized a huge population influx from outside in order to tap the resources of the North. The incoming population's growth in the Murmansk Region throughout the twentieth century was staggering. In 1895, the Indigenous population's share of the region had been 22.3 percent.[8] It had become 0.2 percent by 1989, with an overall population in the Murmansk Region of almost one million.[9] Throughout the twentieth century, the Sámi population remained relatively stable at around 1,600 to 1,900 people. As a result, by the end of the century, the Murmansk Region became the Arctic region with the lowest share of Indigenous people, while being colonized, industrialized, urbanized, and militarized to an extent unmatched anywhere else in the Arctic.[10]

Between the late 1950s and the early 1970s, the policy of *ukrupnenie* (variously translated as "consolidation," "agglomeration," or "amalgamation") was the main force responsible for relocating people in rural areas all over the country. *Ukrupnenie* was a hastily planned, centralized economic rationalization of the whole agricultural sector of the Soviet economy. In the 1960s, central agencies sent ministerial bureaucrats out into the countryside equipped with guidelines to identify "viable" (*perspektivnye*) and "nonviable" (*neperspektivnye*) settlements.[11] Many Indigenous settlements ceased to exist due to these developments. There are estimates that the Soviet Union relocated 70 to 80 percent of all Sámi adults living within the Kola Peninsula, moving them from smaller Indigenous settlements to larger villages or towns.[12] The result was that, in the entire country, some 120,000 villages were deemed worthy of development, against an immense number of 580,000 villages scheduled for liquidation.[13] This one-to-five ratio also applies approximately to the Kola Peninsula.[14] For Indigenous peoples it meant that most of their tiny settlements were closed down, and people had to relocate to bigger settlements.

Additionally, in the Arctic Indigenous regions, the 1957 resolution "On the Measures for Further Economic and Cultural Development of the Peoples of the North" urged "further improvement of the economic and administrative territorial organization of native districts and counties . . . in order to simplify the structure and improve the work of the economic, Soviet and

Party organization in the North."[15] In practice, "simplifying" and "improving" meant dismantling scattered settlements, relocating people to more centralized settlements, and organizing education in boarding schools in order to obtain streamlined structures that would make the local population more legible and controllable and thus fit into the broader Soviet experiment of comprehensive social reorganization.

These measures led to a concentration of Arctic Indigenous people into urban-like model villages from which rural economic activity would spread out into the tundra and bring its fruits back to civilization by means of shift work. The expectation by central planners was that if rural people were offered the amenities of modern life, they would eagerly give up their previous, supposedly miserable, existence. In this volume, Hossein Ayazi discusses similar model villages and similar assumptions in Liberia as an important tool for modernizing and controlling rural economies and populations. Establishing industry-adjacent villages with regulated educational and economic processes can thus be seen as a general trait of modern nation-state projects across the capitalism-socialism divide. In Russian Lapland, the village of Lovozero became this urbanized-village hub. The villages deemed nonviable were closed down. In the Murmansk Region this amounted to the full sedentarization of its Indigenous population. The relocations, however, rarely played out as the Soviet rulers originally intended, leading to consequences that were beyond the imagination of both the relocated people and the Soviet social engineers who designed them. Thus, between 1960 and 1980, after the bulk of these policies had been implemented, life expectancy among all northern Indigenous people dropped by twenty years on average, to about forty-five years for men and fifty-five years for women, mainly due to widespread alcohol abuse, resulting accidents, and suicide. Such behavior, in turn, was a result of chronic housing shortages, changed professional requirements, the separation of families by the means of boarding schools, and a decline of the social status of Indigenous livelihoods and languages.[16]

Schooling Indigenous Children in the Soviet Arctic: The Case of the Sámi

Boarding schools across the Soviet Arctic have played a crucial role in the consolidation and relocation policies described in the previous section. Indeed, "almost all indigenous people in the Russian North remember the boarding school as an important experience in their childhood but with varying evaluations."[17] On the Kola Peninsula, after all relocations had been carried out, the tundra-connected, non-urban population was served in Lovozero mainly by the so-called Native Boarding School (in Russian: *natsional'naia shkola-internat*), a school category that existed across all regions of the Soviet Union with a non-Russian native population. Subsequently, in the Soviet Arctic, these schools, despite their name, did not offer any tuition in the native languages of the North. In areas where native languages were spoken only by a very small minority, such as the Murmansk Region, these languages were not even taught as a separate subject. Tuition followed the general countrywide curriculum for middle and high schools, formally giving pupils the same chances to further their education as in any other regular school of the Soviet Union. The only "native," as it were, element of the schools was the fact that they were organized as boarding schools. Folk dances, singing, or handicrafts were offered as facultative extracurricular activity based on local initiatives.[18] Limited facultative courses of Sámi language were reintroduced from the early 1980s, based on a local initiative, after a twenty-year break.[19] However, the damage created by depriving an entire generation of its native language has been irreversible so far, and native language remains on the decline.

The chronic housing shortage after the relocations of Sámi communities to one central settlement, Lovozero, led to overcrowded flats and generally difficult material and psychological circumstances for the relocated families. Sending the children of relocated families to the boarding school was a consequence of these hardships: an easy way for the local authorities to—apparently—alleviate these problems by massaging the housing statistics.[20] I call this phenomenon the "boarding school paradox," as the parents of boarding school pupils lived in the same settlement, while boarding schools were originally designed to host children whose parents lived in remote settlements without schools. Liubov' Vatonena, a former statistician and one of the first Sámi ethnopolitical activists in the late 1980s, was the first to both locally and globally denounce in public the connection between the housing shortage and boarding school policies: "It was easier to take away the kids than to give housing."[21]

On the one hand, the boarding school was a resource for relocated parents, offering shelter, food, and clothes during a period of material hardship after the relocations. On the other hand, the removal of children tended to increase psychological distress, substance abuse, and social stigma.[22] Clearly, for all my interviewees it would have been preferable to be given the promised housing on time instead of being offered "help" in the form of removing their children from overcrowded temporary dwellings: "When we [the rest of the interviewee's family] moved to Lovozero [from Voron'e; relocation 12, Table 3] we lived . . . they gave us a wooden hut. . . . There were two families living there: my parents' family and my mother's sister's family. The smallest kids lived with them, of course. But we, all the others, studied and lived at the boarding school because with only two small rooms [for both families] there was simply not enough space."[23]

School statistics confirmed this picture. For example, for the 1974–1975 school year, the list of pupils at the Native Boarding School in Lovozero shows 160 Sámi, 57 Komi, 6 Nenets, and 13 Russian children from remote settlements—despite the fact that all remote Sámi settlements had been already eliminated and all families of those Sámi children resided in Lovozero.[24] These ethnic proportions reflect the housing shortage among relocated Sámi families. Lovozero had more inhabitants of Komi descent than of Sámi

Figure 9.1: One of the buildings of the native boarding school with a pencil-shaped playground, probably 1970s, photographer unknown, courtesy of the Lovozero Museum archive.

Figure 9.2: Inside the Lovozero Native Boarding School, probably 1970s, photographer unknown, courtesy of the Lovozero Museum archive.

Figure 9.3: First visit by a Norwegian Sámi delegation at the native boarding school, 1961, photographer unknown, courtesy of the Lovozero Museum archive.

Figure 9.4: In the native boarding school's dormitory, probably 1970s, photographer unknown, courtesy of the Lovozero Museum archive.

descent, and yet there were far fewer Komi children living at the boarding school; they tended to go to the regular daytime school, as there was no pressing need to solve any housing problems for them. The numbers are similar for every school year of that period until the mid-1980s, when more apartment blocks were finally built, gradually relieving the housing shortage.

Besides glossing over the chronic housing shortage, the boarding schools fulfilled several other goals in line with the large-scale and long-term Soviet project of social homogenization. The schools pursued an affirmative action approach, as we would define it today: the active effort to improve life opportunities for members of groups that are in one way or another considered disadvantaged. Providing opportunities to Indigenous children to climb the social ladder within an officially egalitarian Soviet society went, however, hand in hand with suppressing ethnic identity, and with social descent for some. The 1950s onward saw "a dramatic shift away from any attempt at native language instruction in the schools, and use of any language other than Russian was punished."[25] I will now discuss the negative outcomes of this post-relocation, affirmative-action-cum-Russification approach.

Ethnocentrism, Paternalism, and an Implicit Ethnic Hierarchy

The Soviet boarding schools established from the 1950s onward were designed to contribute to the goal of national homogenization. Sovietizing the population amounted to "russifying" the youth both culturally and linguistically. This occurred most intensely between the 1950s and 1970s, affecting one entire generation of Arctic Indigenous people. The schools thus became the main agents responsible for a missing link in language transmission not only within school curricula, but also within families.[26] In addition to a focus on Russian language in education, there were many circumstances that put pressure on the Sámi languages—without a special intention to do so, but rather as a side effect. Although Indigenous languages were not explicitly forbidden in the boarding schools, interview materials clearly show that even in conversations among pupils the teaching staff tried to minimize the use of the Sámi language.

The dominance of Russian was a utilitarian instrument for pursuing the ideological and economic goals of a unitary Soviet people. This dominance slowly grew throughout the Soviet years. While in the early Soviet Union the stated goal was to organize schooling in the native language for each of the over one hundred ethnic groups united in a new communist country, this goal was gradually replaced by a more pragmatic approach. The advantages of a common language became evident to the leadership—when there is a common language, communication is simpler, and people, goods, and ideas travel more easily. The tendency toward uniting people linguistically was for the first time enshrined in law in 1938, when Russian became a mandatory subject at every school of the Soviet Union outside of the Russian-speaking areas. A further boost to the dominance of Russian at non-Russian schools was given by the renewed Soviet overall schooling law of 1958, in which parents were—theoretically—given the right to choose the language of tuition for their children.[27] In practice, this formal right of choice meant two things: First, except for the most populous non-Russian ethnic groups, non-Russian languages were not offered as a tuition language, which meant that for most of the Arctic Indigenous peoples there simply was no choice. Second, many parents and young people came to accept the necessity of a good command of Russian as an indispensable means for success in life under the social circumstances into which they were born. The general attitude among teachers was to convince parents and children of the supposed harmfulness of bilingualism for the child's brain development. Many of them believed in this dominant discourse of the time. These convictions tended to negatively affect the use of Indigenous languages in everyday situations, even at home.[28] Thus, although there was never a decree about the *explicit* suppression of Indigenous languages in schools, in practice, their use was curbed.

The ideological underpinning of this tendency was that Russians were supposedly positioned on the highest level of cultural and economic development within the Soviet house of nations. Thus, despite ideological talk about the eventual merging of ethnic groups into one Soviet people, hierarchical thinking stayed throughout the Soviet era, with the general tendency that the more "native" (meaning: differing from Russian) a group was, the less "developed" it was considered.[29]

The hierarchization of ethnic groups (as a persistent discourse in education), combined with the policy of consolidation (*ukrupnenie*), led, in sum, to strong assimilation pressures and to a heavily reduced Indigenous self-esteem in the new generation growing up mostly in boarding schools. The ethnic hierarchical structures created in the consolidated villages were cemented by the boarding schools hosting the relocated people's children, which fostered ever stronger stereotypes, generally following patterns of social reproduction.[30]

Similar instances of post-relocation social downward mobility due to the proximity of "higher-ranking" ethnic groups have been observed across the entire Soviet North.[31] However, in the case of the Soviet Union it would be wrong to imagine this as a sort of apartheid. In practice, the hierarchy was subverted on a daily basis: interethnic marriage, much easier to realize after relocations, turned out to be the easiest and a very common way to reach the Soviet final goal of ethnic "fusion"; education and career opportunities were in principle open to everyone, and children of different ethnic backgrounds went to school and played together. Constraints in this regard were not formal and absolute, but rather due to structural violence and social dominance, which, nevertheless, were not impermeable.

Destroying Previous Family Structures and Constructing New Gender Roles

Seeing displacement and sedentarization as part of a macroscale social engineering project, namely, forming a new Soviet society, enables us to explain the strong erosion of previous family structures and the establishment of new gender roles. Full sedentarization, among other things, meant that nomadism as a livelihood for the entire family and society had been (*bytovoe kochevanie*) replaced by so-called "production nomadism" (*proizvodstvennoe kochevanie*);[32] that is, reindeer herding was supposed to be done only by paid male herders as shift work, excluding other family members. There was no space for herding and other elements of Indigenous livelihoods in the standardized curriculum. Indigenous livelihood was degraded to a profession for a narrow and purely male part of the population that would have to be learned in early adulthood rather than passed on organically from earliest childhood.

Complete sedentarization and production nomadism had been fully implemented in only a few places in the Russian North, including the lands of the Sámi.[33] The development meant that all Soviet Sámi were living permanently in a settlement—roughly half of them in Lovozero—with some of them, mostly men, periodically leaving for long-term shift work on the tundra.[34] What is more, production nomadism also meant that reindeer herding was defined by planners as the main economic activity of the rural part of Russian Lapland, although this had not been the case earlier. Reindeer monoculture officially replaced the previous polycropping, which among the Eastern Sámi encompassed not only reindeer herding but also fishing, hunting, gathering, and rearing sheep.

This top-down economic reorganization toward monoculture eroded previous family structures. Children were being taken away from families and sent to boarding schools. A strong gender division in professional life resulted from this top-down labor reorganization, and another effect was exogamy, primarily among women, and the resulting widespread bachelorhood of Indigenous men.[35] Where consolidation (*ukrupnenie*), relocation, sedentarization, and a reorganization of labor and education were the main features of social transformation, men tended to suffer a loss of status in society. Women fared better in many cases, with a gain of status thanks to the positively connoted "female" roles they were offered by the majority society in the new urbanized setting. In the Soviet Union, Indigenous women were more likely to find an activity in which they could define themselves in a meaningful way according to discursive categories offered to them by the majority society. These categories of gender positioning were inculcated through education, and this was done most effectively in boarding schools. Indeed, as Konstantinov contends, the new living circumstances "placed ambitious young women on career tracks undreamt of in the context of pre-agglomeration semi-nomadism."[36] Both educational-professional prospects and exogamy also meant that women had a more pronounced tendency to leave their community of origin and move to cities.[37]

The rise of "the educated woman," in combination with the other radical socioeconomic changes, has been a rather consistently realized goal of Soviet power in the Arctic. The new opportunities, however, came at a cost. Young people's ambitions developed in boarding schools, while families were destroyed by forceful separation.

As interviews and archival materials show, boys tended to feel less comfortable with the challenges posed by the boarding school. The following interview excerpts elaborate on this:

> I wouldn't say slow progress [about bad grades among boys]. They were simply not interested at all in studying. Absolutely not interested. They just wanted to go home, to their parents, where they could be in their usual environment, where they could speak their own language, where they didn't feel subject to any prohibitions. But we [two girls], when we came to the boarding school, we decided to study hard. Me and her, we finished [high] school. Already when we were in the seventh grade, we told each other: "Okay, we will study well. We will finish school, and we will get a higher education, no matter what." We kind of swore it to each other. And we did it![38]

The strong gender imbalance in terms of male insurgency and female docility is also reflected in archival materials. Thus, in the second half of the 1970s, among minors caught for law infringements, about 90 percent were boys.[39] There were (and are) of course also boys with ambitions to achieve educational heights. However, tundra work—which requires skills that are not provided in formal education—was conceptualized through the majority society's norms as purely male work, and hence the educational system also pushed young men to be less successful at school and "turn to the tundra."

Vitebsky aptly described the consequences: "Village girls say that they could not imagine marrying a herder, because of their lack of conversation: Television and magazines portray life as a torrent of words and easily revealed passions. . . . To them, the men's communication with animals, rivers and mountains, which make much speech unnecessary, seems maladaptive. Rather than a talent, it is now interpreted as a deficiency."[40] Ulturgasheva offers a convincing approach to understanding this apparently higher "dysfunctionality" of the male part of the population, which resulted in its lower resilience in urbanized settings:

> Unlike the forest, the village is a space which mostly involves human–human (rather than human–animal) relations. Forest challenges and risk situations shape the survival skills and resilience of an adolescent which are compatible with the space of forest; however, even competent forest youths remain more vulnerable in the face of socially risky situations associated with the space of the village (separation from parents, bullying, interpersonal violence, and alcohol abuse). Hence, those who are viewed as resilient in the forest may be vulnerable when facing the necessity of staying in the village, and those who may cope well and show signs of resilience in the village may become quite vulnerable in the forest.[41]

This is a crucial observation. We do not see one gender being resilient, while the other is not. Rather, different genders developed resilience in different areas, depending on the social norms absorbed by them through education and discourses. Many women admit that they avoid the tundra, because they lost their "forest resilience" in their childhood, as these interviewees remark: "But I would like to add, you know what? That my dad was a reindeer herder, he was a reindeer herder, but mum would never go with him to the tundra. And as my mum never went to the tundra, we never lived there. I don't know how to live in the tundra. I never was out in the tundra. And this maybe somehow had an influence on us, I don't know."[42]

> Q: Have you ever thought of going back to the lifestyle of your parents?
>
> A: Oh, no. I've never even thought about it. I wouldn't mind visiting and spending some time where my ancestors used to live. But to work there, clean up after somebody, no thanks. I'm quite squeamish about these things, you know (laughing).[43]

Thus, the gender rift was initiated by the separation of labor and dwelling places after the relocations and resulted in Indigenous male and female lifeworlds drifting apart strongly. This rift was then transmitted to the first post-relocation generation by inculcating the expected gender roles through educational, professional, and cultural norms in the new settlement. In the climate of general post-relocation social despondency, those men who had been relocated as adults soon died away.[44] In the next generation, the mortality among men again was significantly higher than among women. This mortality is mainly due to increased substance abuse, because the gender roles boys were taught since school age left men without prestige and many of them without spouses. The bachelorhood of herders affected not only Sámi men, but also Komi men, and those of mixed origin who engaged in herding after the sweeping social reorganization.[45] However, even before any ethnic activism had awakened in the late 1980s, sociologists had recorded particularly high exogamy and bachelorhood rates among the Sámi women and men, respectively.[46] This relates to the especially low social status of relocated Sámi men, as a combination of, on the one hand, the rural professions being generally devalued by the educational system and labor division and, on the other hand, the poor material and mental situation of the relocated groups. The following interviewee names these factors as a reason why women tended toward exogamy:

A: All this [drinking] everywhere, in Varzino we didn't have that. I don't remember that in Varzino or Kanevka [pre-relocation settlements] people would linger around with bottles, we didn't have that. . . . From the young Varzino men of that time, I cannot tell you one who did not lose himself in the bottle. They all went on the bottle. . . . The women turned out to be stronger in this respect.

Q: So, could we say that the most vulnerable part, if we make such a conclusion, the most vulnerable part of the population became the men among the relocated Sámi.

A: Yes. That's how it was. Here you can't think of or add anything to that. I can't tell you about anyone [men] who would have married here, about any weddings. The girls yes, they married.

Q: Others, not their own men?

A: Well yes, in the sense that they moved here [to Lovozero] and married other men. If those drink like that, who will want to marry a drunk? Drunkards, that's what they said. "Bugger off, you drunkard."[47]

Indeed, between 1975 and 1979, of all marriages among Sámi people, 82.1 percent were mixed. In these mixed marriages, the woman was Sámi in

79.5 percent of the cases, and in the other 20.5 percent, the man was Sámi.[48] The same problem can be put in more general terms: Among all northern Indigenous people, significantly fewer men than women get married, due to the preference for exogamy among women.[49] The low incidence of marriage among men contributed to their social ills and higher mortality as negative psychosocial effects. All in all, the boarding schools contributed a great deal to cementing this new social order, from which mainly Indigenous men emerged as disadvantaged.

Social Reproduction

The third negative side effect of affirmative action in the Soviet educational system for its Indigenous minorities is what Bourdieu and Champagne called the making of "outcasts on the inside."[50] Exclusion within a social system—especially a nominally highly inclusive one like socialism in its idealized form—is more stigmatizing than a priori exclusion from it, because those who are excluded were apparently given "their chance." This means that failure can be more easily attributed to individual deficiency instead of systemic failure, while maintaining the façade of a society of equals. This is a crucial element of the "boarding school paradox" in the formally egalitarian society of the Soviet Union. I subsumed these mechanisms elsewhere under the term "individualization of the negative."[51] The works of Bourdieu, Passeron, and Champagne show how educational exclusion makes a central but concealed contribution to consolidating a given social order in comprehensive schooling systems of modern nation-states.[52] It is therefore especially eye-opening to apply their ideas, originally developed from empirical evidence in France and seen as a social critique of "bourgeois" schooling systems, to the educational system of the Soviet Union.

It appears important to mention the vague but widespread prejudices about the allegedly inborn traits of the Sámi, which were, and partially still are, predominant among the majority population. Such stereotypes offer false explanations for deviant behavior of individuals without connecting this behavior to the systemic social issues that emerged from relocations. Here is another example, from an interview with a non-Sámi and non-relocated person who went to the boarding school in Lovozero and remembers some Sámi children as inherently aggressive.

A: They were like that, aggressive young people; many were aggressive, I don't know why.
Q: You mean specifically among the Sámi kids?
A: Yes, in my class there were many such youngsters. . . . That Vasia, the slightest wrong word, and he could start spitting at you and insulting you all over the place, or he would kick you. There were many such boys. . . . Tolik was also a psycho, and Valia was. . . . They were always so high-strung; you couldn't say the slightest wrong thing to them.[53]

The story goes on, ending tragically with one of those aggressive classmates stabbing a girl to death at a drinking party. The interviewee does not show any awareness that the described aggressiveness could be the effect of social circumstances, while admitting that she does not know why her classmates were aggressive. Aggression was often a mirror of the highly precarious living conditions of the relocated families. It is widely acknowledged today that above-average aggression and violence can be a consequence of social upheaval likely to be transmitted across generations.[54] The high violence and mortality rates among the relocated Sámi groups corroborate this.[55]

My collected corpus of oral history testimonies confirms that widespread despondency among the relocated people—without its causes being understood by the majority population—only reinforced stereotypes connected to the ideologized Soviet hierarchy of ethnic groups. In a vicious circle, misunderstood despondency and despair acted as a social reproduction mechanism, exacerbating the despondency of individuals.

In the Name of Rationality and Progress: The Soviet Union as a "Western" Power

The insights from this chapter allow us to refine understandings of the Soviet Union as a high-modernist state. As we can see across the chapters of this volume, not unlike Christian missionaries in the nineteenth century and corporate-state alliances in the twentieth century in capitalist states, the Soviet Union also facilitated access to basic education for groups regarded as marginalized, above all women or "underdeveloped" ethnic minorities. Soviet power was attracted to the appeal of technical progress and economic rationality, very much like the other, capitalist states situated around the globe.

What all circumpolar countries have in common is that they brought nationhood, state control, intended economic rationality, in short: "progress" and "civilization," to the Arctic regions of our planet during the twentieth century. Based on these insights, I suggest paying more attention to these qualities, which the Soviet Union shared with the "Western" circumpolar states.

Anthropologist Sherry Ortner identifies the roots of what she calls "dark anthropology" in the works by Marx, Durkheim, Weber, Foucault, and Bourdieu.[56] According to Ortner, focusing on the dark sides of human existence has become necessary and unavoidable, with a need first to unravel the colonial past of the West, and then to criticize capitalism, especially in its harsh neoliberal forms.[57] Ortner, however, omits a third important dimension: the criticism of socialist states. Defining the scope of dark anthropology beyond what is typically framed as the dark sides of "the West" gives us a holistic perspective on the dark sides of high-modernist societies, a view independent of their organization as nominally capitalist or socialist. All too often authors slip into seeing the dark sides they analyze as "typical" of one or the other form of polity and social organization, overlooking the commonalities with their "significant other." A strong exception to this is James Scott's book *Seeing Like a State*, in which he elaborates on a range of twentieth-century social experiments across the world, showing their striking commonalities.[58] Among other things, Scott emphasized the "Westernness" of the Soviet Union in the sense of a deeply rooted common denominator between Soviet and "Western" societies: a high-modernist approach to human existence driven by technocracy and simplification (portrayed as rationalism), albeit—according to Scott—with one difference: "For capitalists, simplification must pay."[59] I would qualify this statement by saying that profitability—no matter how (in)efficiently achieved—was at least theoretically an important driver of many Soviet social engineering projects as well. Consolidation (*ukrupnenie*) in its entirety was conceived on the tenets of economic rationalization. Sedentarizing Indigenous people and educating their children in schools was both ideologically and economically driven. In this volume, Hossein Ayazi elaborated on the case of Liberia as a similar pattern, but under the auspices of capitalism and corporate interests. These types of rationalization make the Soviet Union a state more "Western" than is commonly assumed.

Another dimension spanning the capitalist-socialist divide is what I have called the "individualization of the negative." Indeed, blaming the victim is identified by Downing and Garcia-Downing as one of the five fallacies of development-induced displacement.[60] They see it as a recurring pattern, permitting those in power to avoid responsibility. This frequently met attitude exacerbates the negative outcomes of social engineering mega-projects instead of trying to solve them. The blame-the-victim fallacy is perfectly actualized in the "boarding school paradox" discussed in this chapter. Mechanisms of social reproduction through education systems are linked to mechanisms of individualization and exclusion. While Bourdieu and Passeron presented their theory of social reproduction as concepts developed for "Western" societies, they prove applicable to the Soviet education system as well.[61]

In addition to these bridges across the capitalism-socialism divide, many topics related to displacement, boarding schooling, and social reproduction have parallels in research about other groups dominated by a colonizer majority society. For instance, state-induced relocations among the Sámi people have also taken place in Sweden, and the most explicit assimilation policy among all states where Sámi live took place in Norway.[62] In all circumpolar countries, boarding schools for Indigenous children have been a common feature, with amazingly similar recollections among former pupils.[63] The use of boarding schools as a tool to keep children separated from the influence of their families and local cultures has been reinforced in this volume in the case study about Liberia. As was the case in Soviet reality, due to the gender stereotypes proffered by the state, girls were embraced by the system of urbanized upbringing and thus turned out to be more compliant with the system. In the case of the Sámi, the top-down redefinition of gender roles fueled the gradual withdrawal of the women "from the land."[64] This trend has similarly been investigated for Arctic Indigenous peoples outside of Russia—for instance, in Finland, Alaska, and Greenland.[65] In this volume, several cases show how, under the ideal of equality, girls were included in industrial schooling, which had multiple effects, such as fostering equality in a modern "Western" sense, but at the same time undermining functioning Indigenous societal systems and creating new forms of inequality. Thus, for instance, the case described in this chapter resonates with attempts in nineteenth-century North America to confine men to herding and women to hearth and home.

Conclusion

The schooling situation among the Sámi (and other Indigenous people) of the Soviet Union between the 1960s and 1980s was characterized by a peculiar and contradictory tension: On the one hand, we can see straightforward cultural assimilation and social leveling by applying the same centrally planned curriculum across the entire country, regardless of cultural backgrounds. The "good intentions" of this approach were to give the same opportunities to everyone, to unify the country, to converge its ethnic heterogeneity, and to make the country run ideologically and economically smoothly—a process that has been aptly called a domestic Soviet precursor to globalization.[66] On the other hand, the "industrial" dimension of this approach consisted of a loss of Indigenous livelihoods, knowledge, and languages, and a resulting stigmatization of native culture and social reproduction among Indigenous people. The underlying understanding of this civilizing mission was to inculcate a rational understanding of organized, industrial productivity instead of the "disorganized" and "messy" traditional livelihoods.

Similar to the Western powers, the USSR strongly believed in the supremacy of technological progress in the name of rationality. Here is where we find the common denominator of industrial education institutions and corresponding forms of education in the Soviet Union: to bring external understanding of "progress'" to an Indigenous population. While the name "industrial education" was not applied, this chapter shows how the practices were nearly identical to those in the prototypical colonial empires. It is difficult to fully disentangle the aims and pedagogical foundations of Soviet-style colonial boarding schools from those of racialized industrial education.

In conclusion, the common denominator spanning across all twentieth-century Arctic powers was the way in which these nation-states operated as high-modernist powers driven by attempts to teach natives how to live. In the name of rationality and progress, these "civilizing" missions consisted of industrializing traditional livelihoods by introducing a school system focused on inculcating values and knowledge stemming from the majority society. The commonalities of social engineering policies in education and population displacement across the Arctic become manifest in the comparison between the empirical insights from this chapter, other studies in this book, and the existing literature. These commonalities underscore the need for a more holistic view, transcending the widespread socialism-capitalism dichotomy. Truly, as stated in the introduction to this volume, "wherever there was a colony there was likely a set of ideas about how to educate the Natives to work," independently of the underlying religious or economic system.

Thus, concerning their relations with their Indigenous minorities, all circumpolar states of the last century, including the Soviet Union, should be seen,

I argue, as "Western," if we understand the term as a common denominator for a modern nation-state that is governed by a technocratic, industrial rationale of progress through labor division and economic growth. Despite ideological differences, the common industrial dimension of their educational approaches to their Indigenous minorities applies to the Soviet Union as much as to the capitalist countries of the circumpolar world.

Notes

1. Alicia Krömer and Lukas Allemann, "Arktische Assimilierungspolitiken: Indigene Kinder in den Internatsschulen des 20. Jahrhunderts" [Arctic policies of assimilation: Indigenous children in the boarding schools of the 20th century] in *Arktis und Subarktis: Geschichte, Kultur, Gesellschaft* [Arctic and Subarctic: History, culture, society], ed. Gertrude Saxinger, Peter Schweitzer, and Stefan Donecker, Edition Weltregionen (Vienna: New Academic Press, 2016), 103–120; Frank Darnell and Anton Hoëm, *Taken to Extremes: Education in the Far North* (Oslo: Scandinavian University Press, 1996).

2. James C. Scott, *Seeing Like a State: How Certain Schemes to Improve the Human Condition Have Failed* (New Haven: Yale University Press, 1998).

3. The writing of this chapter was supported by the Research Council of Finland under the project WIRE (Fluid Realities of the Wild, 2021–2025, decision no. 342462). Most of the fieldwork was completed within my doctoral research between 2013 and 2018, within the project ORHELIA (Oral History of Empires by Elders in the Arctic, funded by the Research Council of Finland). In Russian Sápmi, I gathered around ninety hours of biographical interviews. These materials were complemented by a large amount of archival materials and local newspaper articles from the Soviet epoch.

4. Tim Ingold, *The Perception of the Environment: Essays on Livelihood, Dwelling and Skill* (London: Routledge, 2000), 6.

5. Ingold, *The Perception of the Environment*, 6.

6. Ulla-Maija Kulonen, Irja Seurujärvi-Kari, and Risto Pulkkinen, eds., *The Saami: A Cultural Encyclopaedia* (Helsinki: Suomalaisen Kirjallisuuden Seura, 2005), 161–164; 261–263.

7. On the Komi and Nenets migration see: Andy Bruno, *The Nature of Soviet Power: An Arctic Environmental History* (New York: Cambridge University Press, 2016), 122–169; Petia Mankova, "The Komi of the Kola Peninsula Within Ethnographic Descriptions and State Policies," *Nationalities Papers* 46, no. 1 (2018): 34–51.

8. Nikolai Aleksandrovich Shavrov, *Kolonizatsiia, eë sovremennoe polozhenie i mery dlia russkogo zaseleniia Murmana* [Colonization, its current situation and measures for the Russian settling of the Murman region], vol. 4 (Sankt-Peterburg: Tipografiia I. Gol'dberga, 1898).

9. "Vsesoiuznaia perepis'naseleniia 1989 goda" [all-union census of the population of 1989], *Demoscope*, www.demoscope.ru/weekly/ssp/rus_nac_89.php.

10. Lukas Allemann, "Soviet-Time Indigenous Displacement on the Kola Peninsula: An Extreme Case of a Common Practice," in *Routledge Handbook of Indigenous Peoples in the Arctic*, ed. Timo Koivurova et al. (New York: Routledge, 2020), 92–105.

11. Neil Melvin, *Soviet Power and the Countryside: Policy Innovation and Institutional Decay* (Basingstoke: Palgrave Macmillan in association with St. Antony's College, Oxford, 2003), 38–72.

12. N. B. Bogdanov, "Protsess urbanizatsii korennogo naseleniia Kol'skogo poluostrova: Saami v XX veke [the process of urbanization of the indigenous population of the Kola Peninsula: the Saami in the 20th century]," *Lovozerskaia Pravda*, November 18, 2000.

13. Melvin, *Soviet Power and the Countryside*, 64.

14. Allemann, "Soviet-Time Indigenous Displacement on the Kola Peninsula"; Tobias Holzlehner, "Engineering Socialism: A History of Village Relocations in Chukotka, Russia," in *Engineering Earth: The Impacts of Megaengineering Projects*, ed. Stanley D. Brunn (Dordrecht: Springer, 2011), 1957–1973.

15. Quoted in Nikolai Vakhtin, *Native Peoples of the Russian Far North* (London: Minority Rights Group International, 1992), 18–19.

16. Vakhtin, *Native Peoples of the Russian Far North*; Yuri Slezkine, *Arctic Mirrors: Russia and the Small Peoples of the North* (Ithaca: Cornell University Press, 1994), 340, 375; Lukas Allemann, "The Sámi in the Spiral of Negative Social Developments of the Soviet North," in *The Sámi World*, ed. Sanna Valkonen et al. (New York: Routledge, 2022), 248–262.

17. Stephan Dudeck, "Oral History of Empires by Elders in the Arctic (ORHELIA): A Common History, a Common Economy, Common Language Roots, and Different Practices Among Four Arctic Indigenous Peoples," in *Oral History: Dialogue with Society*, ed. Ieva Garda-Rozenberga (Riga: Institute of Philosophy and Sociology, University of Latvia, 2013), 72, http://issuu.com/lufsi/docs/oral_history/1.

18. AE, Female, born in the mid-1930s, Lovozero, November 4, 2013. All interview quotations in this chapter have been translated from Russian into English by the author.

19. AA, Female, born in the early 1930s, Lovozero, May 12, 2014.

20. AA, Female, 2014; Anna Afanasyeva, "Boarding School Education of the Sami People in Soviet Union (1935–1989): Experiences of Three Generations" (PhD diss., University of Tromsø, 2019), https://munin.uit.no/handle/10037/15101.

21. Liubov' Vatonena, "Soviet Union: The Problem of the Soviet Saami," *IWGIA (International Work Group for Indigenous Affairs) Newsletter*, no. 59 (1989): 89–91.

22. Lukas Allemann, "'I Do Not Know If Mum Knew What Was Going On': Social Reproduction in Boarding Schools in Soviet Lapland," *Acta Borealia* 35, no. 2 (2018): 115–142; Lukas Allemann, "The Perspective of Former Pupils: Indigenous Children and Boarding Schools on the Kola Peninsula, 1960s to 1980s," in *Sámi Educational History in a Comparative International Perspective*, ed. Otso Kortekangas et al. (Basingstoke: Palgrave MacMillan, 2019), 83–104; Afanasyeva, "Boarding School Education," 79, 186, 191–92.

23. AS, Female, born in the mid-1940s, Revda, July 31, 2014.

24. f.146 op.5 d.194 l.126–132, "Spisok uchashchikhsia v lovozerskoi shkole-internat na 1974–75 uchebnyi god, s ukazaniem natsional'nosti" [List of students in the Lovozero boarding school for the 1974–75 academic year, indicating ethnic belonging], 1974, State Archive of the Murmansk Region, Kirovsk.

25. Lenore A. Grenoble, *Language Policy in the Soviet Union* (Dordrecht: Kluwer Academic Publishers, 2003), 173.

26. Allemann, "'I Do Not Know If Mum Knew What Was Going On'"; Afanasyeva, "Boarding School Education."

27. A. E. Mestnikova, "Etnolingvisticheskoe obrazovanie korennykh narodov: istoriko-pravovye aspekty iazykovoi politiki Rossiiskoi Federatsii XX–XXI vekov" [Ethnolinguistic education of Indigenous peoples: Historical and legal aspects of the language policy of the Russian Federation in the 20th and 21st centuries], *Fundamental'nye Issledovaniia* 2 (2015): 2028–2032.

28. Liarskaia, "Severnye internaty i transformatsiia traditsionnoi kul'tury (na primere Nentsev Iamala)" [Northern boarding schools and the transformation of traditional culture (on the example of the Nenets of Yamal)]; Allemann, "'I Do Not Know If Mum Knew What Was Going On'"; Afanasyeva, "Boarding School Education."

29. Slezkine, *Arctic Mirrors*, 144–147; 343–348.

30. Allemann, "'I Do Not Know If Mum Knew What Was Going On'"; Pierre Bourdieu and Jean-Claude Passeron, *Reproduction in Education, Society and Culture* (London: Sage Publications, 1990).

31. Alexia Bloch, *Red Ties and Residential Schools: Indigenous Siberians in a Post-Soviet State* (Philadelphia: University of Pennsylvania Press, 2004); Igor Krupnik, Igor and Mikhail Chlenov, "The End of 'Eskimo Land': Yupik Relocation in Chukotka, 1958–1959," *Études/Inuit/Studies* 31 (1–2): 59–81. Nikolai Ssorin-Chaikov, *The Social Life of the State in Subarctic Siberia* (Stanford: Stanford University Press, 2003); Bruce Grant, *In the Soviet House of Culture: A Century of Perestroikas* (Princeton: Princeton University Press, 1995).

32. Slezkine, *Arctic Mirrors*, 341.

33. D. D. Bogoiavlenskii, "Vliianie etnicheskikh protsessov na dinamiku chislennosti narodnostei Severa" [The influence of ethnic processes on the dynamics of the number of nationalities of the North], in *Sotsial'no-ekonomicheskie i demograficheskie problemy zaversheniia perevoda kochevogo naseleniia na osedlyi obraz zhizni* [Socioeconomic and demographic problems of completing the transfer of the nomadic population to a sedentary way of life], ed. B. M. Levin, S. N. Batulin, and F. S. Donskoi (Apatity: Akademiia nauk SSSR/Murmanskii oblastnoi ispolnitel'nyi komitet, 1985), 92–93; Kerstin Eidlitz Kuoljok, *The Revolution in the North: Soviet Ethnography and Nationality Policy* (Stockholm: Almqvist & Wiksell, 1985), 127.

34. Bogoiavlenskii, "Vliianie etnicheskikh protsessov na dinamiku chislennosti narodnostei Severa [The influence of ethnic processes on the dynamics of the number of nationalities of the North]," 93.

35. Lukas Allemann, "Yesterday's Memories, Today's Discourses: The Struggle of the Russian Sámi to Construct a Meaningful Past," *Arctic Anthropology* 54, no. 1 (January 2017): 1–21.

36. Yulian Konstantinov, *Conversations with Power: Soviet and Post-Soviet Developments in the Reindeer Husbandry Part of the Kola Peninsula* (Uppsala: Acta Universitatis Upsaliensis, 2015), 162.

37. Allemann, "Yesterday's Memories, Today's Discourses."

38. AS, Female, born in the mid-1940s, Revda, July 31, 2014.

39. f.146 op.5 d.215 l.6–7, 24–28, 53–54, "Otchety o rabote komissii po delam nesovershennoletnikh pri Lovozerskom Raiispolkome 1974–1978" [Reports on the work of the Commission on Under-Age Affairs at the Lovozero District Executive Committee], 1978, State Archive of the Murmansk Region, Kirovsk.

40. Piers Vitebsky, "From Materfamilias to Dinner-Lady: The Administrative Destruction of the Reindeer Herder's Family Life," *Anthropology of East Europe Review* 28, no. 2 (2010): 43.

41. Olga Ulturgasheva, "Attaining Khinem: Challenges, Coping Strategies and Resilience Among Eveny Adolescents in Northeastern Siberia," *Transcultural Psychiatry* 51, no. 5 (October 2014): 632–650.

42. AS, Female, born in the mid-1940s, Revda.

43. VA, Female, born in the late 1950s, Lovozero, September 11, 2013.

44. Lukas Allemann, "The Experience of Displacement and Social Engineering in Kola Saami Oral Histories," *Acta Electronica Universitatis Lapponiensis* (2020): 164–172.

45. Konstantinov, *Conversations with Power*, 234.

46. V. V. Dobrov, V. P. Toichkina, and A. D. Korchak, "Sotsial'no-ekonomicheskie i demograficheskie problemy vosproizvodstva Saamskogo naseleniia [Socioeconomic and demographic problems of the reproduction of the Saami population]," in *Sotsial'no-ekonomicheskie i demograficheskie problemy zaversheniia perevoda kochevogo naseleniia na osedlyi obraz zhizni* [Socioeconomic and demographic problems of completing the transfer of the nomadic population to a sedentary way of life], ed. B. M. Levin, S. N. Batulin, and F. S. Donskoi (Apatity: Akademiia nauk SSSR/Murmanskii oblastnoi ispolnitel'nyi komitet, 1985), 97–98.

47. EK, Female, born in the early 1940s, Lovozero, July 27, 2014.

48. Dobrov, Toichkina, and Korchak, "Sotsial'no-ekonomicheskie i demograficheskie problemy vosproizvodstva Saamskogo naseleniia" [Socioeconomic and demographic problems of the reproduction of the Saami population], 98.

49. D. D. Bogoyavlinsky, "Peoples of Russia's North: Demographic Information," in *Anxious North: Indigenous Peoples in Soviet and Post-Soviet Russia: Selected Documents, Letters, and Articles*, ed. Alexander Pika, Jens Dahl, and Inge Larsen, IWGIA Document no. 82 (Copenhagen: International Work Group for Indigenous Affairs, 1996), 37–38.

50. Pierre Bourdieu and Patrick Champagne, "Outcasts on the Inside," in *The Weight of the World: Social Suffering in Contemporary Society*, ed. Pierre Bourdieu (Cambridge: Polity Press, 1999), 421–426.

51. Allemann, "'I Do Not Know If Mum Knew What Was Going On.'"

52. Bourdieu and Champagne, "Outcasts on the Inside"; Bourdieu and Passeron, *Reproduction in Education, Society and Culture.*

53. GP, Female, born in the early 1960s, August 4, 2015.

54. Judy Atkinson, Jeff Nelson, and Caroline Atkinson, "Trauma, Transgenerational Transfer and Effects on Community Wellbeing," in *Working Together: Aboriginal and Torres Strait Islander Mental Health and Wellbeing Principles and Practice*, ed. Nola Purdie, Pat Dudgeon, and Roz Walker (Canberra: Australian Government Department of Health and Ageing, 2010), 135–144; Amy Bombay, Kim Matheson, and Hymie Anisman, "Intergenerational Trauma: Convergence of Multiple Processes Among First Nations Peoples in Canada," *International Journal of Indigenous Health* 5, no. 3 (2009): 6–42.

55. A. I. Kozlov and D. D. Bogoiavlenskii, "Smertnost' ot vneshnikh prichin i otkloniaiushcheesia povedenie" [Mortality from external causes and deviant behavior], in *Kol'skie Saamy v meniaiushchemsia mire* [The Kola Saami in a changing world], ed. A. I. Kozlov, D. V. Lisitsyn, and M. A. Kozlova (Moskva: Institut naslediia, 2008), 78–85.

56. Sherry B. Ortner, "Dark Anthropology and Its Others: Theory Since the Eighties," *HAU: Journal of Ethnographic Theory* 6, no. 1 (2016): 47–73.

57. Ortner, "Dark Anthropology and Its Others."

58. Scott, *Seeing Like a State.*

59. Scott, *Seeing Like a State*, 8.

60. Theodore E. Downing and Carmen Garcia-Downing, "Routine and Dissonant Cultures: A Theory About the Psycho-Socio-Cultural Disruptions of Involuntary Displacement and Ways to Mitigate Them Without Inflicting Even More Damage," in *Development and Dispossession:*

The Crisis of Forced Displacement and Resettlement, ed. Anthony Oliver-Smith (Santa Fe: School for Advanced Research Press, 2009), 225–253.

61. Bourdieu and Passeron, *Reproduction in Education, Society and Culture.*

62. Patrik Lantto, "The Consequences of State Intervention: Forced Relocations and Sámi Rights in Sweden, 1919–2012," *Journal of Ethnology and Folkloristics* 8, no. 2 (2014): 53–73; Henry Minde, "Assimilation of the Sami—Implementation and Consequences," *Acta Borealia: A Nordic Journal of Circumpolar Societies* 20, no. 2 (2003): 121–146.

63. Krömer and Allemann, "Arktische Assimilierungspolitiken: Indigene Kinder in den Internatsschulen des 20. Jahrhunderts [Arctic Policies of Assimilation: Indigenous Children in the Boarding Schools of the 20th Century]."

64. Piers Vitebsky, "Withdrawing from the Land: Social and Spiritual Crisis in the Indigenous Russian Arctic," in *Postsocialism: Ideals, Ideologies and Practices in Eurasia*, ed. C. M. Hann (London: Routledge, 2002), 180–195.

65. Rauna Kuokkanen, "Indigenous Women in Traditional Economies: The Case of Sámi Reindeer Herding," *Signs: Journal of Women in Culture and Society* 34, no. 3 (March 2009): 499–504; Lawrence C. Hamilton and Carole L. Seyfrit, "Coming Out of the Country: Community Size and Gender Balance Among Alaskan Natives," *Arctic Anthropology* 31, no. 1 (1994): 16–25; Lawrence C. Hamilton et al., "Outmigration and Gender Balance in Greenland," *Arctic Anthropology* 33, no. 1 (1996): 89–97.

66. A. G. Kiselev and S. V. Onina, "Natsional'naia—Nerusskaia—Russkaia: Shkola Iugry v 40–50-Kh godakh XX veka [Native—non-Russian—Russian: The school of Yugra in the 40–50s of the twentieth century]," *Vestnik Ugrovedeniia* 9, no. 3 (2019): 577–587.

CHAPTER 10

Technical Petro-Education and Capitalism in Qatar

Danya Al-Saleh

Introduction

The Company House, a museum in Doha's downtown, details the early years of Qatar's oil and gas industry. The museum is located in the former headquarters of Petroleum Development Qatar Ltd, the first oil company in Qatar and a subsidiary of the Anglo-Iranian Oil Company (which eventually became BP, or British Petroleum).[1] The museum centers the manual labor of Qataris, represented through reenactments of workers' stories and their embodiment as statues (see Figure 10.1), in relation to imported extractive technology and resource wealth. A plaque directly addresses the significance of technological expertise during this period: "Although the West provided the technology required for its search and extraction, oil was secured as much through the hard work and commitment of the Company's Qatari workforce."[2] This tension, of valorizing Qatari labor in relation to Western expertise and oil, resurfaces throughout the museum.

This tension is mediated by quotations, decorating the museum's entrance and exit, that underscore the fundamental value of Qatari labor: "Our human capabilities represent the true wealth and not the oil." These words, from the former Emir, Hamad bin Khalifa Al Thani, along with a similar quotation from his wife, Sheikha Moza bint Nasser, frame the history of Qatari labor's incorporation into the oil industry in relation to Qatar's contemporary development agenda to transition into a knowledge economy. The museum

Figure 10.1: Statues of Qatari workers in the Company House Museum.
Source: Author.

asserts the primacy of Qataris over the nation's subterranean networks of oil and gas and imported fixed capital.

The museum's narrative of Qatari labor as an enduring store of national wealth subtly alludes to the ruling family's current development agenda to transition away from oil and gas. The Qatar National Vision 2030 (QNV 2030), made public in 2008, outlines this agenda to diversify Qatar's economy through a future knowledge economy.[3] This vision aims to prepare Qatari citizens not only to navigate global capitalism, but to produce innovations that can be marketed elsewhere.[4] Qatar's pivot reflects development policies projected by international financial institutions, U.S. think tanks, including RAND, and foreign consulting firms. To build a knowledge economy, Qatari institutions have imported U.S., Canadian, and European universities and colleges, along with investing in a science and technology hub for start-up enterprises.[5]

In this context, I turn to the story of an educational space originating in the oil era. Industrial education in the broader Arabian Peninsula was introduced in the 1930s through the advent of the oil industry. Within Qatar, industrial education was implemented to reinforce British imperial hegemony by managing regional labor unrest through the distribution of skills across racial and ethno-nationalist lines. As this broader volume argues, industrial education is a flexible pedagogical project with various afterlives, and this chapter examines how, over the course of nearly a century, industrial education for semi-skilled manual labor in Qatar's oil industry has evolved into what is now internationally known as technical and vocational education, or TVET. This successor form of industrial education fuses higher education with corporate in-house programs to train Qataris to take the place of "foreign" technicians—manual workers who are categorized as semi-skilled and skilled in the oil industry.[6]

This chapter contributes to the concept of technical petro-education as a lens through which to trouble TVET as a neutral contemporary project where technical skills are distributed. Rather, TVET is a new consolidation of attempts to reform the oil industry's midcentury manual training programs as an attractive option for Qataris, and continues to function as a mechanism to manage the labor hierarchies that characterize Qatar's capitalist society. Currently, noncitizens comprise more than 90 percent of the country's workforce.[7] The exploitation of migrant labor primarily traveling from South and Southeast Asia and East and West Africa, and laboring in construction, security, and the oil and gas industry, has been integral to Qatar's development model.[8] In Qatar's labor market, the language of skill, along with nationality, obscures how workers' wages, benefits, and status are arranged along racial and gendered lines.[9] As the contribution from Zahra Babar and Mishal Khan in this volume shows, technical education in Pakistan paradoxically works to designate South Asian labor as unskilled before even arriving in the Gulf. At this conjuncture, Qatar's ruling class is promoting education as a mechanism for producing Qataris who will replace professional experts primarily from Western Europe, Australia, and North America to manage a future knowledge economy. A contradiction undergirding technical petro-education in Qatar is that it is designed to produce entry- and mid-level Qatari technicians for the oil and gas industry. Technical petro-education, therefore, exposes the class struggle underlying Qatar's attempts at economic diversification.

The Case of Qatar's Technician Certificate Program

In this chapter, I examine this contradiction through the case of an industrial trades education program for the national oil and gas industry in Qatar.

This program, called the Technician Certificate Program (TCP), was housed at the College of the North Atlantic-Qatar (CNAQ),[10] a satellite campus of a Canadian community college founded in 2002, and offers applied certificates, diplomas, and undergraduate degrees in business, engineering, health sciences, and information technology. The TCP aims to prepare Qatari high school graduates, particularly men, to work as entry-level technicians for Qatar Petroleum (QP) and its subsidiaries.[11] Since 2006, when the program was transferred from QP to CNAQ, administrators have experimented with various reforms to attract Qatari men to work as technicians.

I begin by contextualizing CNAQ/QP's joint technician training program in relation to the history of industrial education in the Gulf's oil and gas industry. Industrial schooling has historically been a contested space in Qatar, where oil companies, British officials, and the Qatari ruling elite and merchant class obscure questions of class differentiation by managing the boundaries between foreign and local labor.[12] Initially, oil companies mobilized industrial education as a tool to diffuse labor unrest by promoting it as a nationalist project that would paternalistically uplift the local workforce to eventually operate oil infrastructure themselves. However, this promise of education was supplanted by imaginaries of Qatari citizens populating high-level positions that had been open to Westerners and Qatar's ruling elite. Technical and vocational education, designed to replace foreign labor associated with manual work, increasingly fell out of favor among the Qatari population. CNAQ/QP's program, a product of this history, has repackaged technical petro-education for Qatar's knowledge economy.

The Problem of Technical Petro-Education in Qatar

Sheikha Al-Misnad, a Qatari scholar and education policymaker, wrote in 1984: "The problem of technical and vocational education in the Gulf, being more complicated than usually acknowledged, has always presented a real challenge for both national planners and educators in order to yield even minimal returns in terms of national benefits."[13] In this section, I will contextualize the problem of technical petro-education in Qatar. As an educational project that was first wielded to naturalize racialized and ethno-nationalist labor hierarchies and diffuse labor unrest within the Gulf's oil and gas industry, technical petro-education remained wedded to this work as it evolved over the course of the twentieth and twenty-first centuries.

The Company House Museum positions Qatari labor as the nation's greatest source of wealth. Much of the material that the museum draws on comes from the book *With Their Bare Hands: The Story of the Oil Industry in Qatar*, by Nasser Mohammed al-Othman, the son of a Qatari merchant, who worked for UNESCO during the 1960s. Both the museum and book share a focus on the national character of Qatar's workforce between the years leading up to the first shipment of oil in 1949 and the industry's expansion during the 1950s. Qatari workers' narratives, however, diverge in crucial ways. Unlike the museum, the book details workers' resistance, deaths, and injuries.[14] The book also describes the disposability, for British oil company managers, of these Qatari workers, who were eventually laid off for "redundancies" during the 1960s. Although there is an erasure of non-Qatari manual laborers in the museum, they appear in the background of the book's narrative: an Iraqi who operated the generator at the Company House headquarters and an Omani youth who was laid off by a British manager for eating a tomato.[15] Neither the book nor the museum directly addresses the exploitation of enslaved and bonded labor provided through labor-recruiting agents and the ruling elite.[16]

With Their Bare Hands documents the story of on-the-job training prior to the institutionalization of technical petro-education in Qatar. Company management classified "local hires," itself a contested category, as unskilled manual labor.[17] The book describes some stories of promotions, from "boy" (serving managerial staff) to "coolie" (manual laborer) to the "semi-skilled" positions of driller or driver. These pathways were often circumstantial or occurred due to a worker's attempts to learn about machines and extractive technologies. One worker described how the company's first Qatari drivers came to secure those jobs despite British managers designating roles based on racist ideas about Arabs workers' skills: "When the company brought in some really big vehicles, they also brought in some English drivers, thinking that we Arabs would not be able to handle the big trucks. In the event however it was the English who proved incapable of handling these monsters on the rough desert track so the experienced Qatari drivers . . . drove them with us as their deputies."[18]

While British managers categorized local labor as unskilled, Qatari workers recounted their varied roles building the company's infrastructure. Some Qatari workers collaborated with British and American geologists and drillers during the first few years of the industry, and the company drew on their expertise to develop its geological surveys.[19] During an oil well fire in 1946, it was so-called unskilled laborers making critical interventions under the direction of an American oil fire expert, who stood at a safe distance from the flames.[20] As contributions across this volume show, the categories of skill, and their accompanying wage gaps, have long structured capitalism's international division of labor through racist essentialism. The designation of skill by imperialist and colonial powers through industrial education justified racialized labor hierarchies and the differentiation of labor prices across the globe.[21]

These stories of on-the-job training should be situated in relation to the region's rapid shift from an economy based around trade, fishing, pearling, and pastoralism to entirely new relations and means of production. As Rumaihi argued: "In many respects the introduction of the oil economy was a foreign imposition on Arab Gulf societies—it was operated mainly by foreigners, with the production geared largely toward foreign interests."[22] The process of on-the-job training reflected workers' alienation from the industry's technologies and demands on labor. For instance, an oral history described a death that occurred because the worker had labored most of his life at sea: "Bu Faisal was driving a lorry-load of workers to Dukhan. As the lorry sped along there was a strong headwind and one fellow's headdress flew off. Instinctively, the man held his nose and jumped to go after it, exactly as he would have done if he had been at sea. The poor man died unfortunately."[23]

Workers' alienation from the industry's technology was then used to categorize manual workers as unskilled. One worker explained how, for a time, the company paid donkeys more than workers: "Donkeys were used for road levelling work . . . There was a herd of about 220 donkeys which belonged to Ibrahim Buhmaid, Abdullah Mansour, and Ishaq. They were paid 1.5 rupees per donkey per day—in other words the donkeys were paid more than coolies who . . . only received three-quarters of a rupee."[24]

Additionally, Qatar's merchant class, working with members of the Al Thani ruling family, acquired labor contracts with the company, and through this role taxed and deducted fees from workers—many of whom were the poorest of society, indebted, and enslaved.[25]

Formal technical education programs across the Gulf, developed by oil companies and backed by the local ruling class, were developed to curtail labor actions across national and ethnic lines and to contain certain forms of political nationalism and leftist movements.[26] As an example, in Iran, the Anglo-Persian Oil Company invested in formal education as a mechanism to "Persianize" the workforce and deport Sikh workers who were organizing for higher wages and better housing.[27] Technical education also aimed to separate national workers from those deemed to be foreigners through segregated learning spaces and training programs.[28] In Saudi Arabia, Aramco drew on segregated training for this purpose, along with separate educational programs for American children and Saudi children—the former accredited so that graduates could attend U.S. universities and the latter to prepare Saudis for trade training in the company.[29] Oil companies' early experiments with technical petro-education served as a mechanism to defuse labor unrest by further differentiating between local and foreign workers in the industry while appearing to align company activities with national and regional interest.

There are no studies on investments in education made by the Qatar Petroleum Company under the British. It is possible that the company sent some Qataris to study at technical schools across the region. Additionally, there may have been corporate resistance to invest in education due to the small size of Qatar's population, which, in 1949, was estimated at 25,000 to 30,000 people.[30] For example, in a 1958 comparative study of the oil industry, Finnie argued that "Qatar . . . simply does not have a sufficient supply of competent labor even for common jobs in the company. It is estimated that there are only 630 literate Qataris in the entire shaihkdom."[31]

During this period, state institutions in Qatar developed technical education programs. As the educational system in Qatar grew, the Department of Education (established in 1954 and renamed the Ministry of Education in 1962) created a secondary education track of technical and vocational schools during the 1960s.[32] This track consisted of four schools, one of which was the Industrial School. The Industrial School opened in 1956 as an evening school to "retrain unskilled and semi-skilled national workers and technicians for local needs," eventually shifting in 1964 to "full time courses of secondary school level" that required intermediate school certification in general academic education.[33] The program was chronically under-enrolled.[34] In 1962, the Ministry of Education established a new training center "with the objective of providing Qatari workers with industrial craft skills in order to increase their productivity as well as enable them to replace non-Qatari employees in the long run."[35] It was reorganized, in 1970, as the Regional Training Center (RTC) to develop special courses for employees in Qatar's oil and gas industry. Like the Industrial School, however, the RTC had recruitment challenges. In the following section, I will address why these experiments failed.

Technical Petro-Education in the Era of Industrial Expansion

Just as the Company House Museum turns to contemporary agendas to diversify Qatar's economy, *With Their Bare Hands* centers economic diversification during the 1960s and 1970s through the development of heavy industry in steel, fertilizers, and petrochemicals. The state's expanding institutions invested in industrial education during this period, as discussed above. Yet, industrialization also required a large labor force. AlShehabi explains how the local workforce (in Qatar and across the Gulf) presented

> three major problems for the emerging political and economic elite. First of all, the sheer quantity of capital resulting from the enormous oil revenues, and the number of associated investment activities, exceeded the capacity and quality of the available domestic workforce . . . Second, subjugating the local population to the dictates of capital . . . went against the logic of the modernist welfare state that was supposed to pull local people out of the dire economic situation they found themselves in prior to the windfall from the oil revenues. It would have been a hard political sell to place locals under the increasingly hungry and urgent demands of the new-found capital . . . Finally, the reliance on local and Arab workers created . . . a political consciousness among the workforce that demanded a greater say in the distribution of oil revenues.[36]

This non-Arab "foreign" labor force, which could be deported at any moment, corresponded with a new system of citizenship where benefits made possible by oil revenues maintained strict divides between foreign and local labor.[37] Thus, what is commonly described as the Gulf region's demographic "imbalance" emerged in this period.

Formal attempts to implement technical education in Qatar were entangled in this demographic imbalance. Across the Gulf, these tracks could not attract enough youth due to better opportunities in other sectors. Nationals across the region developed negative attitudes toward technical education, which tended not to result in high-level work and was associated with manual, and therefore "unskilled," labor.[38] The Qatari government experimented with categories of skill that the British used to divide and control labor along racial and ethnic lines.[39] As Natasha Iskander has argued, skill in Qatar never actually corresponded to ability in relation to a trade, and instead remains a political category: Foreign laborers "came to be regarded as unskilled, irrespective of their actual ability. The rights of local [Qatari] workers were enlarged to include agency, aspiration, and freedom of contract, and locals were given the status of skilled employees, even when they had little skill at all."[40] Although the objective of technical education was to produce skilled and semi-skilled Qatari workers, this educational track remained associated with manual work, categorized as "unskilled." Above all, scholars and planners were aware that "even if all these [Qatari] students from the secondary schools were sent to post-secondary technical education . . . the total number would still constitute only a very small fraction of the total number of vacancies available in the job market."[41]

Despite this, state institutions, since the 1980s, wedded technical petro-education to the project of nationalizing the workforce, commonly described as Qatarization. The project was first launched within QP, the national oil and gas industry. Managers identified technical petro-education as a space that needed to be reformed to produce more qualified Qatari workers. The evolution of QP's Technician Certificate Program within CNAQ is the product of this project.

CNAQ opened its doors as the country's only technical postsecondary institution in 2002. In 2006, Qatar Petroleum relocated its in-house technician training program to CNAQ, undergoing significant reforms over the following decade. While this program struggled for similar reasons as identified across earlier periods, a novel challenge undergirding CNAQ/QP's program is that it produces entry-level technicians for the oil and gas industry in an era when the country's ruling class is promoting a future knowledge economy where Qataris are envisioned to assume roles as researchers, engineers, entrepreneurs, and doctors.

The following section identifies two attempts to reform this program between 2006 and 2019, revealing contradictions regarding the future of fossil fuels and capitalism in Qatar.[42] The first is how this program continues to be entangled in the project of managing the boundaries between national and "foreign" labor in Qatar's oil and gas industry. The second addresses how reforms to facilitate continued education for the program's graduates clash with the dictates of capital, which requires most of its Qatari technicians to stay fixed in their roles as technicians. Just as the British experiments with technical petro-education were entangled in the emergence of the Qatari nation and its racialized labor hierarchies in the age of fossil fuels, a Canadian satellite campus became strangely wedded to the story of contemporary attempts to mold Qatari labor.

The Qatari Technician Between the Foreigner and National Divide

Technical petro-education has functioned as a tool to obscure national class differentiation by distinguishing Qatari labor from foreign labor. Not only have these categories of labor shifted over time, the question of exactly *what forms* of labor Qatarization efforts are targeting has not always been clear, even to those implementing nationalization programs.[43] Given Qatar's national population size, policymakers have repeatedly argued that it is impossible to nationalize the entirety of Qatar's workforce.[44] Instead, planners determined that resources should be spent on developing a cutting-edge educational system to replace upper-level foreign managerial and technical labor with Qataris.[45] In fact, QP/CNAQ's reforms focused on differentiating entry-level Qatari technicians from non-Qatari technicians by transforming the program from a competency-based education (the Technician Preparatory Program, or TPP) into an internationally certified track (renamed the Technician Certificate Program, or TCP).

The renaming signals the decision in 2006 to transform QP's two-year competency-based training program into a four-year certification.[46] QP management considered international certification a mechanism to overcome educational challenges impeding Qatarization by differentiating Qatari technicians from foreign semi-skilled and skilled labor. These reforms of the TPP into the TCP also centralized the process of recruitment, interviewing, and training to reduce competition between QP and its sister companies in their Qatarization efforts.[47]

Managers and educators understood the program's relocation to CNAQ as a mechanism to manage the relationship between Qatari and foreign labor. A Canadian faculty member pointed out the significance of producing a separate space for Qatari technicians to learn without having to deal with workplace tensions: "when they [Qataris] come to work in a plant everyone is hiding info[rmation] . . . they hide it because they know if they are Qatari, they [the foreign instructors] don't want to train them [Qatari trainees] so they can't replace them [the foreign instructors]."[48] In a study of Qatarization within QP, Al-Horr recounts an interview with a manager who argued that transferring the program to CNAQ ensured that Qatari technicians remained in the company: "Some expatriates used to deal 'harshly' with the national trainees, who had just finished their theoretical training and joined a company for on-the-job training, just to keep them trainees as long as possible, which might make the trainee leave the company, while in the new system, as soon as the trainee finishes the four-year program, he/she will get a job title and a permanent position."[49]

While faculty and administrators reproduced various ideas about foreigners in the workplace, they themselves were unclear on exactly who in the workplace was controlling information. At times, they appeared to be describing other technicians rather than the managers in charge of overseeing technicians' training and educational pathways. The figure of the foreigner in the training process, however, was a broad category describing all workers threatened by Qatarization—a process that implied they would eventually be replaced by Qatari labor.

By relocating TCP outside the workplace, CNAQ administrators and faculty blurred their own role managing the training of Qatari technicians. All TCP students are trainees, paid company employees. Their four-year training counts within QP's ranking system, which positively impacts their salaries and position within the company. In this way, TCP faculty simultaneously function as educators *and* managers of students. Through this dual role, they sometimes serve as company informants, disciplining Qatari technicians by reeducating them, not just on how to operate machines, but also on how to relate to work and authority. A Canadian faculty member described TCP students as exceptionally "underperforming" and "lazy."[50] In an interview with an instructor regarding the early days of the program, he described how faculty struggled with attendance and participation. To deal with this, they developed a system which disciplined students by withholding their salaries if they accumulated too many absences.

In fact, CNAQ faculty and administrators developed their own ideas regarding their role in Qatarization. A Canadian faculty member pointed out the importance for the nationalist project of "open[ing] up opportunities for Qatari nationals to operate and maintain their country's infrastructure." Emphasizing contemporary global hierarchies of knowledge production, he made the case that technical education was more suitable for Qatar: "if I want a design, I go to Siemens . . . same [I go abroad] if I want an aircraft." Elaborating on this point, he argued that "here in this country you need to operate and maintain tech. . . ." Returning to the question of Qatarization, he concluded: "the industry would prefer to have their own Qatari nationals doing this work . . . the only problem is that there is not enough." Rather than framing the program as producing distinguished Qatari technicians for a future knowledge economy, he situated technical petro-education within a global system where Qatar would continue importing Western technology far into the future.

TCP's role obscuring the social nature of class through an unstable national/foreigner divide has even extended to the figure of the Qatari technician. When I was conducting interviews in 2018, the program was struggling to attract enough students. This was exacerbated by competition with the military, particularly after a new law, announced in March 2018, required a year of military service for every Qatari man after high school graduation. The new law made recruitment difficult and led to increasing numbers of Qatari men enlisting with the military. In response, the category of the Qatari technician was stretched to expand the pool of eligible employees. A QP/CNAQ employee described how the program decided to open to "non-nationals" with Qatari mothers.[51] The first student in this category had been admitted the year before. In many ways, TCP's admittance of students with Qatari mothers in this category is strategic since they are often not granted the same educational opportunities as Qatari citizens. The program, in this way, reworks technical petro-education's historic role of obscuring class differentiation in Qatar through the stretching and constriction of nationality.

The question of redefining the category of foreigner was also a discussion point. A faculty member who had taught for nearly a decade at CNAQ explained how the campus had been pushing the government to consider the value of non-Qatari students: "These students have lived here their whole lives and attended government subsidized schools—you have invested in these kids."[52] He continued, describing the value of these students as investments in Qatar's national future: "The massive number of non-Qatari students who are willing to work for cheaper and build the country."

This framing of non-Qatari immigrant communities must be situated in relation to the oil industry's ongoing assessment of how profitable it is to Qatarize its workforce. In a QP study, some managers made the case that replacing Western employees in high-level managerial roles with Qataris would be more profitable in the long term.[53] In the case of replacing low- and mid-level workers with Qatari labor, the company's calculations were less clear. This is because there are guaranteed pay scales and human resource laws for Qatari labor that surpass the salaries and benefits of many non-Qatari technicians.

In this way, technical petro-education remains a contradictory space, working to maintain and manage the divide between foreigners and nationals, while obscuring class differentiation in Qatar. In the next section, I examine a second wave of reforms, starting in 2015, which aimed to make TCP more attractive to Qatar's population by transforming it into an academic credit-based program.

Vertical Permeability and Lifelong Learning

On November 27, 2018, I sat in a lecture by CNAQ's outgoing president, Dr. Ken MacLeod. The lecture focused on the evolving role of TVET in Qatar. In attendance were CNAQ faculty and administrators, representatives from the Ministry of Education and UNESCO Doha, along with policymakers and government officials. Asserting that building a knowledge economy in Qatar would not happen overnight, MacLeod situated CNAQ in relation to Qatar's more prominent investments in higher education, proclaiming that "we are [also] in the human capital business." CNAQ's goal, just like elite institutions such as Georgetown University or Texas A&M's branch campuses in Qatar's Education City, is to also produce "knowledge workers." Imagining TVET for Qatar's future, he described CNAQ's transformation into a twenty-first century polytechnical institute that would eventually include a TVET high school and graduate program. Turning to the role of industry, he argued that the college's business partners "need to see themselves as part of the education system." He took CNAQ's recent collaboration with QP to reform the company's in-house program as an example. Starting in 2015, QP and CNAQ initiated a second wave of reforms, which aimed to further integrate technical petro-education within Qatar's educational system by enabling "vertical permeability" for its graduates.

Vertical permeability entails facilitating pathways from trades and apprenticeship certifications to higher-education diplomas and degrees. Trades programs, even when partnered with postsecondary institutions and accreditation bodies, have historically not provided access to further education. Graduates often face "an impasse that force[s] them to default to high school credentials to initiate a new, separate post-secondary educational journey that did not account for knowledge and skills attained through trades programming."[54] Even with certification, TCP graduates were often unable to pursue further studies in Qatar without starting from scratch. Further reforms, starting in

2015, centered around enabling vertical permeability by awarding academic credit and developing a credit transfer policy to CNAQ's engineering technology diploma program.[55]

Vertical permeability aligns with another buzzword of neoliberal education, lifelong learning—the notion that the economy and its impending transformation requires workers who are willing to continue to upskill—adding value to themselves by constantly pursuing new skills and knowledge. Through educational reforms, the future Qatari technician would do more than merely perform tasks—they would also learn "higher-order thinking skills" to resolve workplace issues. As MacLeod reminded the audience, "all of this is occurring with the backdrop of the rapidly emerging fourth industrial revolution characterized by technological transformation that will fundamentally change the way we live and work, and therefore the way we educate and train."

Vertical permeability is promoted as a fix to Qatar's problem of making technical petro-education attractive in the context of knowledge economy development. Starting in 2015, QP decided to develop a program that awarded academic credit. The process entailed QP consulting companies about their technician workforce needs. The program was also transformed from a four-year into a two-year program with standard phases. Once students complete TCP, they receive a certificate and forty academic credit hours. At the time of my research, CNAQ was also developing a bridge program where students could study another year for a diploma.

Administrators, such as MacLeod, positioned TCP in relation to the broader restructuring of CNAQ, where productive collaboration with industry produces optimal TVET educational opportunities. The stories of students, however, countered this narrative of vertical permeability. While TCP graduates now receive academic credit, graduates continue to be confused about the process of transferring to other programs in Qatar and abroad. The program is accredited through Canada following the Canadian Tech Accreditation Board, which is not acknowledged everywhere or even at many higher-education institutions in Qatar.

Sitting in a computer lab at CNAQ, I met three students for an interview.[56] While they were all currently pursuing a Process Automation diploma, two were graduates of QP/CNAQ's TCP program during the early 2010s. Halfway through the interview, one of them turned to me, flipping our roles: "So now we have a question for you . . . So the Texas [A&M University at Qatar] and Education City you are from, working there, right?"[57] After I responded affirming this, the student continued: "So if you want to transfer our degree, our certificate to Texas, what is the procedure and how many years does it take?" Uncertain, I responded based on my knowledge of Qatar's broader educational landscape: "I'm not sure, so I think the case would have to be, what [body] is this program accredited by, and does it transfer to a U.S. institution? So, it seems like from what I've been learning . . . is that this program . . . it is accredited by a . . . Canadian board . . . And that board . . . tends to be more equivalent to . . . UK [United Kingdom] programs . . . I don't know [whether] Texas A&M would take your credits, because it depends if your credits are transferable." The student concluded: "So I have to start from the beginning."

Discussing their different options to pursue bachelor's degrees, both TCP graduates emphasized that their decision-making centered on the time it would take to acquire a higher degree. One graduate explained why he considered studying in a less prestigious program: "See how old my kids [are] now and I'm still in diploma . . . a university in UK . . . I forget the name of the university, they came out to Qatar and did some program and I just give them my certificate and I give them my . . . experience . . . and they said: 'Okay I will give you a very good option . . . I am accepting all your grades and credits, plus your skill experience for five to six years and also your certificate. I will give you a Master's [degree].'"[58]

After having spent six years working on their education and training, their main concern was achieving the promise the reforms of TCP were making: vertical permeability. This, however, was not a seamless process within Qatar's higher-education landscape.

A few months later, I interviewed Ahmed and Majid, two TCP graduates who were pursuing a CNAQ diploma. Ahmed explained how he had graduated from QP/CNAQ's program in the early 2010s and returned to the campus five years later to work on a diploma. He told me: "I wish I didn't do TPP."[59] After high school, he had tried to study at a U.S. branch campus in Education City but found it too challenging. He explained that when he was approached by QP he had no idea he was signing up to be a technician. He immediately justified this from the perspective of the industry: "Look it makes sense. The country brings in workers to do these jobs and they have to provide everything for them on top of salary [health, benefits] and they need Qataris to do this." Ahmed's disappointment with his career path had to do with treatment on the job. Ahmed demonstrated the problem for me through role-play with Majid: "look, imagine you're an engineer and I'm the technician. If you're in the room no one will even look at me. Like I'm nothing and don't know anything."

As CNAQ reforms itself to provide technical petro-education, not as a lesser track of education, students themselves experience their education as lesser, attempting to continue their studies to acquire promotions and respect. Their continued education, however, depends on approval by their manager. Without it, they might be forced to pay back the tuition and salary granted to them when they were trainees. Majid proudly told me that he was the only person from his division who got permission to continue his studies. Recounting the story of how he successfully requested to pursue a diploma, he described how his boss challenged him: "They ask me to give one reason to sign this paper to let me study. I said: 'I want your position.'" He also told his manager how the diploma would teach him how to properly maintain their company's machinery, describing how "foreign" technicians "bypass this procedure which is harmful for the machine." "It's not his country," Majid reasoned, "they don't care about the machine."

Ahmed described why he was pursuing a diploma. He explained: "I came here because I don't understand what's going on in the plant. Before they agreed to send me to study here, the paper [I had] for my position is a high school diploma." Describing the amount of time that he had already spent studying, and how this did not translate into respect, he told me: "You know my father [says] all the time, he's like: you're thirty-one. When will you finish with school?" Ahmed wanted to eventually become an engineer. He explained: "the problem with [TCP] is that they didn't . . . teach us to be an engineer. I applied . . . to the UK to take a degree. [My company] didn't accept and they told me you . . . have to be a technician for four years."

Transferability functions as a contemporary fix to the contradictions underpinning the project of technical petro-education in Qatar. These reforms reflect broader shifts in strategies to Qatarize technician positions in the oil and gas industry. Transferability is intended to shift social understandings of technical petro-education as a lower educational track. Despite reforms where TCP graduates accumulate academic credit, these credits only transfer to certain institutions. Additionally, pursuing further education is contingent on receiving their employer's approval. A diploma did not necessarily translate into changed working conditions either; throughout the interview, Ahmed and Majid described friends who had acquired diplomas and were doing unrelated work in their company. In this way, these reforms of technical petro-education promise a future of lifelong learning, while fixing Qatari labor in place.

Conclusion

Technical petro-education is a contradictory space in which the struggle between capital and labor underlying Qatar's contemporary attempts to transition away from oil and gas appears more visible. During the first half of the twentieth century, technical petro-education was entangled in attempts to manage labor unrest. Oil companies strategically promoted technical petro-education as nationalist projects that would uplift the local workforce to eventually operate fossil fuel infrastructures themselves. Starting in the early 1960s, the Qatari state's experiments with technical petro-education were entangled in ongoing struggles to manage the relationship between national and foreign labor—whose exploitation currently remains the foundation of the Gulf's political economy. As the promise of education was interwoven with imaginaries of Qataris populating high-level positions, technical petro-education consistently struggled to attract Qatari students. CNAQ/QP's program, a product of this history, has reformulated technical petro-education once more to manage labor relations in Qatar. This chapter traced recent attempts to produce more Qatari technicians by either elevating their position above other non-Qatari technicians or promising them a future of lifelong learning—the notion that capital will require workers who continue their education to deal with transformations associated with automation and machine-learning technologies.

Yet, the figure of the Qatari technician continues to trouble Qatar's spaces of higher education. In an interview about outreach challenges at Texas A&M's engineering branch campus, the problem of technical petro-education emerged unexpectedly. Describing a meeting the Texas A&M employee had with a CEO of a QP subsidiary, the American employee summarized the executive's grievance: "Technicians and workers want promotions, but this is not what the 'organization needs.' The problem is that the company needs 75 percent technicians and 25 percent engineers." In this sense, reforms of technical petro-education do not align with the dictates of capital. Companies want to keep certain Qatari workers in low- and mid-level technical positions. This is even more so the case in multinational oil and gas companies that have less pressure to implement Qatarization and prefer to hire labor from abroad who have limited benefits and legal protections.

The project of producing Qatari technicians exposes the fragility of class differentiation in Qatar's capitalist society. The foreign worker, who comprises approximately 90 percent of the workforce in Qatar, is propped up as a foil that the Qatari population must eventually replace. CNAQ's attempts to repackage technical petro-education are challenged by the ways that students' aspirations are shaped by the promises of Qatar's national development agenda, which emphasizes building an educational system that prepares Qataris to compete within a post-oil international order. The ideal products of Qatar's educational system are not promoted as entry-level technicians, but upper-level managers, researchers, and engineers. As the U.S. employee, recounting the Qatari CEO's analysis of Qatarization, made clear, capital does not need its Qatari technicians, such as Majid and Ahmed, to transform themselves into engineers by continuing their education. It is not profitable.

Notes

1. Qatar became a British protectorate in 1916 and gained formal independence in 1971.

2. These words gesture to the role of technological expertise mediating the relationship between Gulf rulers and the United States and Britain during the twentieth century in the Arabian Peninsula. See: Michael Christopher Low, "Desert Dreams of Drinking the Sea, Consumed by the Cold War: Transnational Flows of Desalination and Energy from the Pacific to the Persian Gulf," *Environment and History* 26, no. 2 (2020); Toby Craig Jones, *Desert Kingdom: How Oil and Water Forged Modern Saudi Arabia* (Cambridge: Harvard University Press, 2010); Natalie Koch, *Arid Empire: The Entangled Fates of Arizona and Arabia* (London: Verso, 2023).

3. General Secretariat for Development Planning, "Qatar National Vision 2030," 2008, www.psa.gov.qa/en/qnv1/Documents/QNV2030_English_v2.pdf.

4. For a critical analysis of the QNV 2030 and discussion of how these reforms were implemented, see: Alī Khalīfa Al-Kuwārī, "The Visions and Strategies of the GCC Countries from the Perspective of Reforms: The Case of Qatar," *Contemporary Arab Affairs* 5, no. 1 (2012): 86–106.

5. Education City, an urban development project, which hosts six prestigious U.S. universities, is promoted as the center of knowledge economy development in the country. For studies of Education City, see: Danya Al-Saleh and Neha Vora, "Contestations of Imperial Citizenship: Student Protest and Organizing in Qatar's Education City," *International Journal of Middle East Studies* 52, no. 4 (2020): 733–739; Jana Kleibert et al., "Transnational Education Zones: Towards an Urban Political Economy of 'Education Cities,'" *Urban Studies* 58, no. 14 (2020): 2845–2862; Neha Vora, *Teach for Arabia: American Universities, Liberalism and Transnational Qatar* (Stanford: Stanford University Press, 2018).

6. I am drawing on the categories of labor that are used by recruitment agencies for the oil and gas industry across the region. See: Andrea Wright, *Between Dreams and Ghosts: Indian Migration and Middle Eastern Oil* (Stanford: Stanford University Press, 2021).

7. "More than ninety countries were represented among Qatar's two million foreigners in 2017." See: Natasha Iskander, *Does Skill Make Us Human? Migrant Workers in 21st-Century Qatar and Beyond* (Princeton University Press, 2021), 17.

8. Abdulhadi Khalaf, Omar AlShehabi, and Adam Hanieh, eds., *Transit States: Labour, Migration and Citizenship in the Gulf* (London: Pluto Press, 2015).

9. Natasha Iskander's study of construction workers in Qatar argues that skill is a political category, rather than a marker of technical competence, to exclude, exploit, and control those designated as "unskilled": Iskander, *Does Skill Make Us Human?*; See also: Wright, *Between Dreams and Ghosts*; Amélie Le Renard, *Western Privilege: Work, Intimacy, and Postcolonial Hierarchies in Dubai* (Stanford: Stanford University Press, 2021); Michelle Buckley, "Construction Work, 'Bachelor' Builders and the Intersectional Politics of Urbanization in Dubai," in *Transit States: Labour, Migration and Citizenship in the Gulf*, ed. Abdulhadi Khalaf, Omar AlShehabi, and Adam Hanieh (London: Pluto Press, 2015), 132–149.

10. In February 2022, CNAQ was nationalized and renamed the University of Doha for Science and Technology. Leading up to this, the status of CNAQ's relationship to College of the North Atlantic was unclear. According to the interviews I conducted with CNAQ administrators between 2018 and 2020, the college was not a branch campus, but rather operated as a consultancy model providing management services and academic integrity. By 2019, CNAQ's operations were fully taken over by national bodies in Qatar, with CNA only involved in accreditation. I adopt the term "satellite campus," which was used by most CNAQ administrators at the time of my research.

11. Khalid Al-Horr, "Political Contingency and the Implementation of Localisation Policies: A Case Study of the Oil and Gas Sector in Qatar" (PhD diss., University of Leeds, 2009), 230.

12. Michael Edward Dobe, "A Long Slow Tutelage in Western Ways of Work: Industrial Education and the Containment of Nationalism in Anglo-Iranian and Aramco, 1923–1964" (PhD diss., Rutgers University, 2008); Hassan Mohammad Abdulla Saleh, "Labor, Nationalism, and Imperialism in Eastern Arabia: Britain, the Shaikhs and the Gulf Oil Workers in Bahrain, Kuwait, and Qatar, 1932–1956" (PhD diss., University of Michigan, 1991).

13. Sheikha Abdulla Al-Misnad, "The Development of Modern Education in Bahrain, Kuwait and Qatar with Special Reference to the Education of Women and Their Position in Modern Gulf Society" (PhD diss., Durham University: 1984), 197.

14. The book, however, does not address strikes during this period. For accounts of labor organizing across the Arabian Peninsula during this period, see: Jill Crystal, *Oil and Politics in the Gulf: Rulers and Merchants in Kuwait and Qatar* (Cambridge: Cambridge University Press, 1995); Iskander, *Does Skill Make Us Human?*; Andrea Wright, "Migratory Pipelines: Labor and Oil in the Arabian Sea" (PhD diss., University of Michigan, 2015); Robert Vitalis, *America's Kingdom: Mythmaking on the Saudi Oil Frontier* (Stanford: Stanford University Press, 2007).

15. Eating the tomato was a violation of company policy. Fresh produce was provided only to British employees. Nasir Muhammad Al-Othman, *With Their Bare Hands: The Story of the Oil Industry in Qatar* (London: Longman), 73.

16. "British diplomatic reports on the oil company's operations indicate . . . in 1949, at least 250 men out of the 2,220 Qataris on its roll were known to be enslaved, hired out to the company at the personal request of the enslaver, who often as not was a member of the royal family." Iskander, *Does Skill Make Us Human?* 46.

17. See Iskander, *Does Skill Make Us Human?*

18. Al-Othman, *With Their Bare Hands*, 93.

19. Iskander, *Does Skill Make Us Human?* 46.

20. Al-Othman, *With Their Bare Hands*, 65–67.

21. As Walter Rodney argued, "the international division of labor brought about by imperialism and colonialism insured that there would be the maximum increase in the level of skills in the capitalist nations." Walter Rodney, *How Europe Underdeveloped Africa* (London: Verso, 2018), 213.

22. Mohamed G. Rumaihi, "The Mode of Production in the Arab Gulf Before the Discovery of Oil," in *Social and Economic Development in the Arab Gulf*, ed. Tim Niblock (New York: Routledge, 1980), 49.

23. Al-Othman, *With Their Bare Hands*, 55.

24. Al-Othman, *With Their Bare Hands*, 52.

25. Michael Field, *The Merchants: The Big Business Families of Saudi Arabia and the Gulf States* (London: J. Murray, 1984).

26. Dobe, "A Long Slow Tutelage."

27. Dobe, "A Long Slow Tutelage."

28. Dobe, "A Long Slow Tutelage," 34.

29. Dobe, "A Long Slow Tutelage," 91, 95.

30. Ahmed Mohamed Marzooki, "The Development of Vocational and Technical Education in Qatar" (PhD diss., Durham University, 1994), 19; Dobe, "A Long Slow Tutelage," 42.

31. David H. Finnie, *Desert Enterprise: The Middle East Oil Industry in Its Local Environment* (Cambridge: Harvard University Press, 1958), 119–120.

32. Abdulla Juma Kobaisi, "The Development of Education in Qatar, 1950–1977 with an Analysis of Some Educational Problems" (PhD diss., Durham University, 1979).

33. Al-Misned, "The Development of Modern Education," 229.

34. Kobaisi, "The Development of Education in Qatar," 65–66, 78.

35. Kobaisi, "The Development of Education in Qatar," 99.

36. Omar AlShehabi, "Histories of Migration to the Gulf," in *Transit States: Labour, Migration, and Citizenship in the Gulf*, ed. Abdulhadi Khalaf, Omar AlShehabi, and Adam Hanieh (London: Pluto Press, 2015), 14–15.

37. AlShehabi, "Histories of Migration to the Gulf."

38. Al-Misnad, "The Development of Modern Education," 215.

39. Iskander, *Does Skill Make Us Human?*

40. Iskander, *Does Skill Make Us Human?*, 53.

41. Al-Misnad, "The Development of Modern Education," 217.

42. I draw on research conducted between 2018 and 2020, including interviews with administrators, faculty, and students at CNAQ and other institutions in Qatar.

43. When QP announced that companies should aim for 50 percent Qatarization in 2000, managers diverged on how to calculate nationalization. Ultimately, QP decided to calculate Qatarization in relation to employees . . . who are categorized as "skilled" and "semi-skilled," removing "unskilled" and "temporary" labor from the overall workforce numbers. See: Al-Horr, "Political Contingency and Implementation of Localization Policies."

44. Al-Misnad, "The Development of Modern Education," 215–216.

45. Al-Misnad, "The Development of Modern Education," 203.

46. TCP was certified by Technical and Further Education in accordance with the Australian National Training Authority standard. Al-Horr, "Political Contingency and Implementation of Localization Policies," 231.

47. Al-Horr, "Political Contingency and Implementation of Localization Policies," 236–237.

48. Interview, November 2018.

49. Al-Horr, "Political Contingency and Implementation of Localization Policies," 236–237.

50. Interview, November 2018.

51. Children of Qatari mothers with non-Qatari fathers are not granted citizenship at birth. Additionally, there is a population of Qataris without citizenship status or with differentiated citizenship rights determined according to when their families' migration to Qatar was recorded. Across the Gulf there is a population who do not have citizenship or were stripped of citizenship, known as *bidoon*.

52. Interview, October 2018.

53. Al-Horr, "Political Contingency and Implementation of Localization Policies."

54. Gamar, "Vertical Permeability," 209

55. Gamar, "Vertical Permeability."

56. Interview, September 2018.

57. At the time, I was a visiting researcher at Texas A&M University at Qatar, an engineering branch campus.

58. Interview, January 2019.

59. TCP was called TPP when he was in the program. In our conversation, he would slip between TPP and TCP.

CHAPTER 11

The Paradox of Technical Education and the Pakistani Migrant in the Gulf

Zahra R. Babar and Mishal Khan

Introduction

Nestled on prime commercial real estate in the bustling city center of Rawalpindi, Pakistan, sits an impressive six-story building. The Alliance Manpower company (Alliance),[1] one of Pakistan's largest overseas employment promoters (OEPs), occupies all six floors of the building, where visitors are met with a hive of activity. One of the premier "one-stop" shops for Pakistani workers seeking employment overseas, Alliance attracts hopeful migrants from across the country to its doors.

Although business is good, Alliance, like many OEPs in Pakistan, faces a steady barrage of complaints from its most important clients—employers in Gulf states.[2] Time and time again, Alliance representatives are told that Pakistani migrant workers do not have the skills needed for the jobs they are hired to perform.[3] This complaint is not only expressed by employers; the Pakistani state has also adopted this narrative. In 2015, Pakistani provincial labor minister Raja Ashfaq Sarwar led a delegation to Doha with the goal of expanding Pakistan's supply of workers to Qatar. He stated: "It is not just the number; it is the quality that we want to send from Pakistan. We do not want to send just the routine type . . . We are personally going to supervise and we are going to send the right people for the right job."[4] The Pakistani state's goal, expressed through its rhetoric, publications, and public announcements, is to ensure that its citizens are well equipped for overseas employment: educated with the skills needed not only to contribute to but to succeed in the booming Gulf economy.

Alliance has thus aligned its mission with national policy by not only providing brokering services but also offering state-of-the-art technical and vocational training to all its Gulf-bound workers. Workers have also heeded this message. Many migrants spend large sums of money to obtain technical credentials, motivated to take any steps necessary to maximize their chances of success. International organizations such as the International Labour Organisation (ILO) also emphasize technical education as a critical policy tool to ameliorate the negative experiences that South Asian workers have when they migrate to the Gulf for work.[5] However, despite these investments in creating a modern high-skilled labor force—by the state, international organizations, private actors, and workers themselves—Pakistani workers continue to be channeled into low-wage occupations such as construction and transportation, subject to mistreatment, or sent back to Pakistan.

In this chapter, we are faced with an altogether different context from the nineteenth-century industrial education schools that once served as sites of discipline, punishment, and moral reform of the working poor. We focus not on children being instilled with the gospel of work and money, but rather adult men who are all too attuned to the reality that contributing to the capitalist economy through wage labor and grueling work is the only way to survive. Indeed, over the decades, droves of Pakistani workers have migrated for the chance to participate in the global labor market. With its rhetoric of upskilling and meritocracy, we argue that technical and vocational education is primarily a site of *performance* rather than education. We focus less on the role that technical education plays in orienting racialized workers to the will of the global economy, and more on its primary importance as a site in which the myths of capitalism are rehearsed.

We highlight the paradoxes generated when state actors attempt to craft their rhetoric around skill, job readiness, and capability in the face of a labor market that is indifferent to these traits but rather seeks endless streams of replaceable manual labor. Gulf employers require labor that can be carried out by anyone for a short period of time, while restrictive state immigration bureaucracies ensure that workers are cycled back to their home country. Why then do individuals, the Pakistani state, and the private sector continue to invest in technical education?

The elaborate technical education apparatus, of which Alliance is but one actor, with its emphasis on arming migrants with the skills needed to succeed in the Gulf, allows the Pakistani state to maintain several illusions. The first illusion is that the Pakistani state is an equal player in the Gulf-Pakistan economic relationship. This rhetoric masks the fact that contemporary labor migration is a crucial development strategy for a Pakistani state entirely dependent on the Gulf labor market both to absorb Pakistan's surplus working population and to be a reliable source of remittances. The need to protect this labor corridor has propelled the Pakistani state to allow Gulf states to encroach on its own sovereignty.

The second illusion is that technical education programs facilitate access to both social mobility and professional advancement for Pakistani workers. The Pakistani state thus reproduces the neoliberal claim that participation in the global economy is determined by skill and credentials, and that the capitalist system is one that rewards merit rather than ascribed characteristics such as gender, race, and ethnicity. Many of the challenges that Pakistani workers face are attributed to the fact that workers lack skill. However, as Natasha Iskander has argued, "unskilled" as a descriptor does not "map onto actual observed ability" but instead marks specific racialized bodies as "unable to acquire skill."[6] This contradiction—investing in the supply of highly skilled labor even while Gulf employers demand unskilled and cheap labor—actively facilitates the accumulation of capital in the Gulf and reinforces racial capitalist structures in the region.

Furthermore, we connect this specific contemporary dynamic to broader histories of racial capitalism and empire—the theme of many chapters in this volume. The racialized labor corridor between the Subcontinent and the Gulf must be anchored within the longue durée and historical arc of colonial infrastructures of labor reallocation.[7] We show how racial ideologies have played a lasting role in shaping ideas around skills, nationality, and race in the Gulf.[8] Thinking in the racial capitalism tradition forces us to examine how differentiation inheres in the very processes that claim the language of skilling, industrial education, and universal opportunities of the free market in the twenty-first century.

A Snapshot of the Phenomenon

Before we begin our historical analysis, it is important to describe key features of the contemporary Gulf-Pakistan labor migration corridor and to sketch the structural challenges workers face. While we heed the warning of scholars not to exceptionalize the Gulf region,[9] the sheer scale of migration from the Subcontinent to the Gulf states is a clear indication of this migration corridor's vital significance to the global economy. While numbers vary, the United Nations and the ILO estimate that in 2019, there were between 23 and 35 million foreign workers based in the six Gulf states, comprising a majority in five out of these six countries' national workforces.[10]

Beginning around the middle of the twentieth century, the six Arab monarchies of Bahrain, Kuwait, Oman, Qatar, the United Arab Emirates (UAE), and Saudi Arabia began attracting millions of migrant workers seeking employment opportunities. Existing scholarship suggests that the Persian Gulf's incremental evolution as the third-largest global hub of international migration occurred due to both pull factors, such as its booming hydrocarbon sector, as and push factors, such as lethargic economic development in many migrants' countries of origin.[11] The discovery of hydrocarbon and subsequent state efforts to exploit these highly lucrative resources generated a demand for labor that could not be met by the relatively small local populations.

Since the 1970s, when the Pakistani state began tracking migration statistics, Saudi Arabia and the UAE have been the two leading destinations for Pakistani workers. Furthermore, the six Gulf monarchies have consistently ranked among the top five migration destinations for Pakistanis globally, with recent estimates indicating that approximately 95 percent of all Pakistani labor migrants are based there. The total number of overseas Pakistanis, including those who are *not* considered labor migrants—i.e., "highly skilled" workers and middle-class professionals—was estimated at 8.4 million in 2021. Approximately 53 percent of the total number of all citizens classified as "Overseas Pakistanis" were working and living in the Gulf countries.[12]

The vast majority of Pakistani workers in the Gulf are single male migrant workers, who populate jobs in the lower tiers of the labor market—most significantly, building and construction, craft and manual work, and transport.[13] The total annual remittances overseas workers send back to Pakistan has averaged around twenty billion U.S. dollars over the past few years—discounting the unusual COVID-19 pandemic year of 2020—with almost six billion coming from Saudi Arabia and another four billion from the UAE.[14] Pakistani migrants based in Saudi Arabia and the UAE provide half of all global remittances that Pakistani migrants send home. As a comparison, Pakistanis based in the United States and the United Kingdom combined remit approximately 2.5 billion dollars annually.[15]

Migration to the Gulf poses a number of risks. Migrants often incur considerable amounts of debt to cover their recruitment and travel fees.[16] Migrants pay large fees for their work visas expecting that their earnings will enable them to rapidly pay off these debts. This is not always the case. Given the tenuous nature of their contracts and visa status, migrants are limited in their ability to navigate the issues they confront after arriving in the host country—including potential job loss, as the layoffs in the early days of the COVID-19 pandemic demonstrated.[17]

Under the *kafala* system—the Gulf version of a guest worker program—migrants are entirely dependent on their employers. Upon returning to their home states, Gulf migrants may face additional barriers imposed by their labor recruiters, who take limited responsibility for their long-term well-being, unwilling to compensate failed migration journeys—they are, after all, a profit-oriented business. Migrant experiences are thus shaped by a variety of factors based on the positionality of workers in both Pakistan as well as in the Gulf.

Perhaps most significantly, the Pakistani state relies heavily on remittances from migrant laborers. So much so that facilitating the flow of migrant workers is considered a crucial development strategy for the national economy. The state, therefore, has a vested interest in ensuring that Pakistani workers remain desirable to Gulf labor employers. Pakistan's regulatory environment is constantly modified to ensure the smooth running of this vital migration channel. International agencies often put pressure on the Pakistani state to comply with international standards that promote safe and transparent migration regimes. However, the Pakistani state also has very little control over the practical bureaucratic mechanics of migration. It is in fact Gulf state and private actors who oversee worker health and medical screenings, skills vetting, and the collection of biodata for identity authentication processes.

Pakistani officials justify these intrusions on their sovereignty by framing them as a form of burden-sharing that strengthens the relationship between sending and receiving states—a mutually beneficial collaboration. However, in material terms the relationship is characterized by marked inequality, as these intrusions serve the economic interests of Gulf states. Pakistan has emerged as one important example of a global phenomenon: the commercialization of labor migration enabled by a state-driven agenda to market its citizenry for overseas employment.[18]

The structural conditions described here provide vital insight into the broader political and economic context within which the Pakistani technical education apparatus is embedded. Given these constraints, a worker ending up in line to work at a construction site (or being sent back to Pakistan altogether) rather than being able to acquire work as a mechanic, technician, or plumber has very little to do with a worker's actual possession of technical skills. Rather, we must look further afield to the structure of the Gulf-Pakistan relationship and the racialized nature of the labor market in the Gulf.

Capitalism in the Gulf: Centering Race and Empire

In recent years, scholars have been translating the insights of Black Marxism to contexts beyond the Americas and Europe.[19] We, like others in this volume, insist that paying attention to the simultaneity of racialization and capital expansion provides powerful insights into the ideological dimensions of industrial education projects across time and place. Further, examining the crucial role that South Asian migrants play in Gulf economies—by quite literally building these centers of finance, extraction, and exchange with their own hands—is an opportunity to study how race and capital accumulation reinforce each other from the vantage of an increasingly central node in the global economy.

The Gulf states are experiencing growth and financial expansion at an accelerated pace, and the crucial role the Gulf plays in the circulation of the global flows of capital and finance makes this region key to understanding global capitalism and its "spatio-temporal dynamics."[20] While the region should be understood as a distinct geographical space of accumulation, it must also be seen as deeply imbricated in European and American economic and imperial interests.[21] We insist that colonial racial hierarchies remain key to understanding how the Gulf labor market functions today. The pace and scale of growth experienced in the Gulf has been spurred on by assigning distinct roles to racial groups across the spectrum of elite and working class, skilled and unskilled, professional and manual work.

That racism shapes the lived experiences of migrant workers is clearly articulated by workers themselves. As one informant related about his experience in Saudi: "A dog is worth more to them than us. When they see a Pakistani, they spit on him and say that 'you have destroyed our country.'"[22] This kind of everyday symbolic violence is a mundane part of the South Asian migrant experience in the Gulf states. Much is known anecdotally and through numerous ethnographic accounts about the crude stereotypes that define the labor market in the Gulf—the tireless Filipino domestic worker, the rebellious Arab, and the lazy African migrant.[23] The tangible impact these racial tropes have on the labor market is clearly demarcated not only in popular discourse but in official visa categories that vary by nationality.

These negative attitudes toward migrants have traditionally been explained by the ethnocratic constitution of the Gulf monarchies where the valorization of pure nationality based on blood creates biases against all foreigners.[24] Outsiders have thus been excluded or, as scholars note, economically incorporated but juridically kept in a state of permanent transience.[25] This is so even as foreigners are an absolute majority in most Gulf states.[26] However, there is a growing demand to look beyond the citizen versus noncitizen divide in the Gulf by explicitly interrogating how processes of racialization structure inequality in significant ways.[27] Foreigners are, after all, a racially variegated group with hierarchically organized relationships to resources and status.

Furthermore, when we see the Gulf region as inheriting colonial racial legacies, the contours of the labor market become predictable.[28] The racial ideologies that permeate the Gulf labor market cannot simply be understood as biases integral to "Arab" cultural practices or the persistence of premodern forms of tribal hierarchy. Attention must be given to the reconfigurations and specifically modern forms of racial essentialism that facilitated the spread of global capital under the aegis of European empires. For example, Amélie le Renard examines how categories of "White," "Western," and "skilled" work are co-constituted in Dubai.[29] Renard notes that the racialized labor categories that exist today in the Gulf states emerged from their distinct status as British protectorates. These dynamics had a deep impact on the privileged access that "White" British and European individuals enjoyed across the region—a privilege that has extended into the contemporary period.

Similarly, it was within Britain's nineteenth-century global empire that the category of the "Indian" laborer was born. Stereotypes about racial groups and their proclivities, or lack thereof, to labor were rampant.[30] The British contrasted East Asian so-called coolies—who were seen as industrious and hardworking, but always a threat to the social order—with Indian labor, whom they viewed as docile and accustomed to service. Connected to this perception was the view, according to nineteenth-century British philosophers, that India was "over-stocked with labor but vacant of capital and skill."[31] Meanwhile, British abolitionists, liberal imperial agents, and early anthropologists insisted that Indian society was naturally harmonious and hierarchical.[32] But throughout the nineteenth century, perhaps what was most important to the British was that India provided a source of cheap yet "free" labor. These ideas about Indian labor legitimized Britain's incorporation of indentured labor within a legal order that increasingly lacked tolerance for enslaved labor after the abolition of slavery in the British empire in 1834.[33]

Maintaining this image of Indian labor required major ideological and institutional investments on the part of the British colonial state. Part of the colonial state's strategy was to deny the existence of slavery on the Subcontinent—a form of abolition by denial or by semantics.[34] These ideologies allowed the colonial state to create bureaucratic structures that facilitated both state-sanctioned and private "free" labor migration across the globe.[35] These streams of labor spread from India, to the Caribbean, to Fiji, to Natal, and along many other international paths forged by colonists.[36]

Links between the Subcontinent and the Gulf were particularly strong in the nineteenth century under formal imperial migration structures—although much migration took place outside these official channels as well. The port of Aden in modern-day Yemen was in fact annexed and officially governed as part of the British Raj, while consulates with a significant degree of political and economic influence were established in the surrounding areas across the Persian Gulf.[37] Scholars have noted the movement of men and children from Makran, on the Baluchi coast, and the wider Subcontinent across the Arabian Sea to work on pearl diving, date, and other industries in the Middle East.[38] Historians caution against relying on the "slave-free" divide to understand this history, as migration took a variety of forms.[39] Andrea Wright has shown how mobilizing Indian labor to work in the oil fields of the Gulf into the early twentieth century drew explicitly on labor recruiting patterns and infrastructures—such as the Bureau of Emigration—already present in India during the colonial era.[40]

Understanding the contemporary role that Pakistani labor, in particular, and South Asian labor, in general, play in the Gulf labor market demands that we attend to these historical dynamics. The provocation of the racial capitalism literature is not merely that racial hierarchy is "baked into the structure" of modernity and capital expansion, but rather it invites us to see the "ongoingness" and intensification of processes of racialization under the guise of modernity, equality, and market neutrality.[41] The Subcontinent has provided vital labor for the global economy since the nineteenth century and these racializations persist, even while dominant institutional structures in both the Gulf and Pakistan tout the language of merit, multicultural liberalism, and skill.

The Pakistani state's economic, political, and psychological investments in technical education thus mischaracterize the root cause of the challenges faced by Pakistan migrant workers. Instead of seeing the persistent devaluation of South Asian labor and skill as a structural dynamic with roots in the nineteenth century, actors on both sides of the labor market corridor attribute the problem to a deficit in individual skill. Like many of the industrial education projects covered in this book, the modern technical education apparatus is meant to maintain and legitimize rather than challenge the systemic forces that racially segregate the labor market.

Between the Gulf and South Asia: The Twentieth Century

We now turn our attention to political and economic events of the postwar era. While official imperial structures were dismantled after the formal end of the British Empire, the Gulf remained a key site of resource extraction for the global economy dictated by British and ultimately American capital.

During the early phases of oil extraction—the 1940s in Saudi Arabia and Kuwait, and the 1950s and 1960s in other Gulf states—the Gulf states turned to their Arab neighbors to meet their growing labor needs.[42] Egyptians, Yemenis, Sudanese, Syrians, Lebanese, and Palestinians flocked to the Gulf by the thousands during the first wave of oil-driven economic growth, and soon began to establish large expatriate communities.[43] Arab migrants to the Gulf worked in every sector of the region's expanding economy. These migrants numerically dominated jobs available not only at the lower tiers of the labor market, such the construction and service sectors, but also in "skilled" occupations such as banking, engineering, medicine, and education.[44] There was a brief period during which Indians were recruited as "skilled" labor alternatives to regional labor by British corporations—for example, under the Anglo-Persian Oil Company in Qatar—but this was short-lived and many Indians ended up working in low-wage work once again.[45]

By providing their labor to critical sectors of the economy, many Arab migrants were able to repeatedly extend their visas and work contracts, until ultimately they became de facto long-term residents. These workers were able to bring their families with them and remain in Gulf countries for decades despite their temporary legal status and the absence of paths to citizenship. Over time, these communities became a visibly multigenerational and semipermanent feature of the Gulf's demographic landscape, provoking sharp concern and anxiety from state actors, who feared that non-Gulf Arabs could easily assimilate with local citizen populations, infect them with Pan-Arabist and leftist radical ideologies, and drive demands for social, economic, and political change.[46] Gulf rulers wanted a sustained flow of foreign workers to be rapidly and easily deployed to meet various labor shortages, without becoming permanent settlers. Since the beginning of the oil era, Gulf leaders consistently described the flow of migrant labor as a temporary phenomenon tied to particular and necessary development goals. Retrospectively, Gulf rulers and policymakers may well have wished for tighter control over the early inward flow of their Arab neighbors.

Over time, the "Arab" worker—a category that included labor from a variety of regions around the Middle East—emerged as a politically problematic figure. An alternative was needed. Driven by political exigencies, social concerns, as well as the economic demand for the cheapest labor available, from the 1970s and even more forcefully from the early 1990s, the pattern of Gulf migration and its ethno-national composition shifted eastward.[47] This change was undertaken deliberately and strategically, as more and more South and Southeast Asian migrants arrived to replace earlier cohorts of Arabs. Not only were Asian laborers willing to work for lower wages than non-Gulf Arabs, their cultural, linguistic, and religious differences would also make them less likely to integrate with local citizen populations. Scholars have noted how essentializing differences between Arabs and Asians has served to undermine South Asian claims of belonging to Gulf.[48] Alongside the shift to a non-Arab workforce, the Gulf states also engaged in more focused policy efforts to strictly limit migrants' ability to permanently settle. Gulf leaders refined their arsenal of policies to govern and control migrant labor, made blood lineage a prerequisite for access to citizenship, and restricted opportunities for social and cultural integration between their citizenries and the foreigners who increasingly outnumbered them.[49]

Just as most other parts of the world were at a stage of rolling back, reshaping, or completely dismantling their temporary guest-worker programs, the Gulf states were actively doing the opposite: strengthening and solidifying the *kafala*.[50] This system enabled the Gulf monarchies to cycle large numbers of migrants in and out of the region on short-duration contracts, fulfilling the never-ending labor needs of capitalist enterprises. The Gulf's developmental goals and economic ambitions were thus completely delinked from the available labor pool of Gulf citizens. At the same time, new tools were locked in place to reinforce the "temporariness" of labor migrants' stay and to ensure their eventual return home upon completion of their contracts.

By the early 2000s, the Arab expatriate population across the Gulf had decreased significantly, and the regional labor market as well as the population as a whole became dominated by Asian migrants.[51] While this turn to

Asia as a labor source has been justified as occurring on the basis of assorted economic logics, it is clear that the shift was not prompted by the sudden lack of availability of the non-Gulf Arab worker population. This shift also did not occur because Arab migrants lacked the requisite skills, expertise, and educational levels needed by Gulf employers. If anything, unemployment pressures in many of the large-population Arab states such as Egypt, Syria, and Jordan have significantly increased over the past decades, as have the educational levels of their populations. It is the capacity for the Gulf states to ensure political and social exclusion of their migrant population and the historical role that Indian labor has played across various centers of production and capital accumulation that are the root causes of this shift away from Arabs to South Asian workers in particular.

Returning to the Paradox: The Technical Education Apparatus

The Gulf's mid-twentieth-century demand for Asian—particularly South Asian—labor to replace Arab workers drew on racial imaginings and imperial structures of previous centuries. Yet today, rising economic powers in the Gulf such as the UAE have staked their global reputation on claiming a new modern and cosmopolitan identity, espousing a neoliberal vision of equal opportunity, individual achievement, and mobility. Both receiving and sending states thus promote the narrative that modern education and proper training is all that stands in the way of individuals securing a better future for themselves and the extensive families they often support.

Despite these promises, however, technical education has not delivered transformative outcomes for the Pakistani working class, either at home or abroad. In fact, technical education was originally a (failed) development strategy within Pakistan decades before labor migration became the major source of rural and urban livelihood that it is now. During the late 1980s and early 1990s, the dominant donor-driven narrative prompted the growth of semi-public "rural support" programs (RSPs). These were designed to not only address the immediate effects of poverty through infrastructural investments and social sector programs in healthcare and education, but to also broaden "livelihoods" goals through enterprise development, microfinance, and skills creation.[52] These RSPs promoted development initiatives across a variety of sectors to aid struggling rural or semirural communities, promising people vibrant economic and social conditions in which to thrive.

Vocational educational offerings for the rural poor in the 1980s and 1990s were designed to tie closely to the needs of the local labor market, to address local skills gaps, and to develop technical capacities that would match people with local employment opportunities. The aim was to support pathways for income generation for people in the places where they lived, and to enhance local capacity so that Pakistani citizens did not have to migrate. Implicitly, these programs were meant to discourage domestic migration, alleviating the pressure on Pakistan's already over-stretched urban centers. By the 1990s and early 2000s, however, the reality of low capital investment and the crisis of failed structural adjustment policies in Pakistan altered the motivation and orientation of the Pakistani labor market. Neither urban nor rural economies could accommodate the millions of people who needed to earn a living. The future of work for Pakistan's laboring population was to be found abroad.

Despite the fact that earning opportunities are exponentially higher in the Gulf, Pakistani workers confront formidable obstacles when trying to obtain positions outside of low-skilled industries. As discussed, Gulf employers frequently invoke a "mismatch" between migrants and their jobs, which often negatively impacts relations between workers and their employer. Employers justify cutting wages and altering contractual terms not *because* their employee is Pakistani, but ostensibly because a Pakistani worker does not have the requisite experience or skill to carry out assigned tasks. Employers complain that this mismatch forces them to invest either in training workers or reassigning them to alternate tasks and hiring replacement workers to do the actual work stated in the contract.

However, as Iskander and others have argued, the possession of skill is never an objective fact but always an assessment that is inflected by racialized and gendered structures that already assign certain values to different categories of humans—designating some as being even less than human. Gulf employers claim they need "better-quality" workers, situating responsibility squarely on the shoulders of sending nations. These arguments serve to retrench the idea that migrants themselves are to blame for the problems they encounter. Believing that migration is their ticket to upward mobility, Pakistani workers setting off for the Gulf often become trapped in place through a racialized structure that devalues their skill, labor, and personhood.

The Pakistani state is compelled to respond by increasing its efforts to train and upskill workers. To that end, both the state and private bodies have established technical training and vocational education centers in or close to the parts of the country with lower literacy rates, low access to traditional academic education, and limited occupational opportunities. Along with agricultural and industrial education, these "skill-based" educational systems have been redesigned and repurposed over time. Today, they are centrally configured in pre-departure migration processes preparing Pakistani workers to join the global labor market. In 2005, Pakistan established the National Vocational Technical Education Commission to enhance the skill level of Pakistani workers so that they remain "globally competitive" for the Gulf's labor market.[53] Almost four thousand public and private technical and vocational training institutes fall under the jurisdiction of the Commission, and about half a million Pakistani students attend their programs each year.[54]

The private sector is also investing heavily in addressing the so-called skills deficit of migrant workers. Returning to present-day Rawalpindi, Alliance Manpower and its physical presence are a testament to this fact. The building boasts a large executive suite, a gleaming wood-paneled boardroom, and an open-plan space on the ground floor where an army of agents is busy on phones and computers.[55] On another floor, there are multiple smaller rooms where human resources personnel—mostly from the Gulf—conduct one-to-one interviews with job seekers looking for opportunities in the Middle East. An entire wing is composed of a large auditorium where, among other activities, Alliance representatives give pre-departure lectures to educate migrants about the social and cultural contexts of the UAE, Saudi Arabia, Qatar, and other Gulf nations.

Alliance tests candidates on a variety of technical skills and proficiencies and matches them with host Gulf employers. Gulf companies send delegated representatives to Pakistan to speak to job candidates and vet their skills credentials. Each floor of Alliance is dedicated to addressing particular aspects of the recruitment process. Most impressive of all are the extensive teaching and training facilities, where plumbers, electricians, masons, and dozens of other "skilled" workers can be rapidly produced, as per the requirements of overseas employers. According to Ahmed, Alliance Manpower's deputy manager,

> This business has been running for the last 17 years, we got our proper license in 2005. Initially, we were doing business only in Saudi. We got visas from different companies who required labor

> and distributed them to agencies over here. The problem this raised was that the agencies here would not provide the perfect people for the company and our clients would be dissatisfied. The companies in Saudi relied on us to meet their demands. If they required a steel fixer, we would convey the demand to agencies over here but when the laborer reached Saudi, he didn't know anything about steel fixing. Hence, the companies would hold us accountable and complain that we were sending them the wrong people. We realized that it will be much better to have our own place dedicated to managing the whole process and decided to open this office.[56]

Ahmed explained that Gulf employers' complaints about skills matching prompted his family to expand their recruitment operations to become a "one-stop" shop. Today, any OEP that wants to succeed in this business has to take an active role in developing the skills of the workers they send overseas. Ahmed admitted that many Pakistanis are desperate for employment. Tahir, another labor recruiter interviewed, explicitly described how attentive the organization is to the varying skill requirements needed for different categories of workers: "There are not many requirements for unskilled labor and we just have to check if the applicant is physically sound. In Pakistan we refer to them (unskilled potential migrants) as class 4 and we just need to check their fitness. For skilled and semi-skilled ones, there are proper documents and certifications that we have to review, and there are local training centers which approve them and their skill level. There are also private training centers approved by the government. If a person comes to us and says he is a plumber and we do not know his skill level, we send him to the training center to check if he is a qualified plumber. This way, no one can lie us about their skill level."[57]

Both Pakistani and Gulf officials largely view OEPs in Pakistan as unethical and profit-oriented actors driven by fees generated from processing migrant visas and contracts. Pakistani and Gulf officials routinely accuse OEPs of sending untrained and untested workers. According to Ahmed, Gulf-based employers now insist that OEPs provide a security deposit for any large job contracts they are awarded, which are retained in escrow. The deposit is released and returned to the recruiting agency only when an OEP demonstrates that the migrant workers it sends have the capacity and skills to properly complete a contract. Hearing this message loud and clear, Alliance decided that it made business sense to directly manage and control the process of vocational and technical training. By investing in the quality of technical education, Alliance attempts to ensure that the workers it sends overseas have undergone extensive training under its close watch.

Further "quality control" measures are taken by Pakistan's Bureau of Emigration and Overseas Employment (BEOE)—commonly known as the Bureau. This is a state apparatus within the larger Ministry of Overseas Pakistanis and Human Resources Development and is tasked with authenticating the various skills claimed by individuals. The Bureau maintains a record of all Pakistani migrants and supervises the operations of OEPs, with the authority to revoke licenses and shut down businesses that defraud migrants. In the presence of deputed government officials, migrants are obliged to sign a Foreign Service Agreement, which includes full details of the terms of employment.[58] There are currently eight regional-level Protectorates of Emigrants Offices that operate under the BEOE, which organize much of the paperwork needed to support the regular annual labor migration of about one million Pakistanis. Most of the Pakistanis who come through the Protectorates' offices are blue-collar labor migrants, the vast majority of whom end up in the Gulf.

The contradictions produced through this uneasy marriage are apparent in the policies and practices of the Bureau. While this institution is formally committed to protecting the rights of its overseas citizens, it is simultaneously driven to ensure that international employment opportunities for the Pakistani labor class continue to grow. These two sets of interests do not align and require a constant balancing act on the part of the Pakistani state. The provincial arms of the Bureau have adopted nomenclature used by the British a hundred and fifty years ago, calling themselves "The Protectorates of Emigrants," and are colloquially—though not always affectionately—referred to as "The Protectorate" by both migrants and labor recruiters alike.[59]

The Bureau oversees all processes of migration for Pakistani citizens in both the pre- and post-departure phases. It is also meant to intervene on behalf of migrants should things go awry. In reality, however, once migrants are in the Gulf, the Bureau struggles to track labor law infringements and contractual noncompliance. The burden of providing support to migrants is left to the over-stretched labor attachés based in Pakistani embassies in the Gulf. There are, however, far too many Pakistani workers and too few labor attachés to effectively address each individual case. Pakistan is also reluctant to engage in direct state-to-state confrontation to demand that Gulf receiving states are accountable for protecting the rights of its citizens.

In addition to the many state and private stakeholders active in Pakistan's labor migration architecture, there are also a number of civil society organizations. These groups are dedicated to addressing the skills deficit issue as part of broader programs to improve the migration experience. According to the Migrant Resource Center (MRC)—a semi-public entity tasked with supporting Pakistani migrants, particularly in the pre-departure phase—equipping migrants with the appropriate skills and qualifications gives them a far higher chance of a "successful deployment." The would-be migrants who approach the MRC for information and advice come from diverse walks of life. Khayyam Sohail of the MRC office in Islamabad explained how skills, both academic and technical, are central to determining what migratory options to discuss with would-be migrants.

> So normally someone would walk in, and say "I am a student, I have just done my Bachelors I want to go abroad." That's one potential scenario. Another scenario is someone who is semi-skilled or unskilled who walks in and says "I am a driver" or "I am a cook, I have been doing this for x, y, z amount of time and I saw that you provide information on migration so where can I go?" And then there will be people who are highly skilled who will say "I want to go to Canada or Australia for immigration so how do I go about doing that?"
> So there are different scenarios, people will come [to the Migrant Resource Center] from different backgrounds. So when someone comes in, we talk to him, we assess what sort of skill set he has, what qualifications, technical education, what knowledge does he possess, and then we advise him [of his options] accordingly. If it is someone who is young and does not have any experience or skill, we can refer those people to the technical and vocational training institutes, which now provide free education in most cases up to a 6-month diploma is free. So they can go upgrade their skills and then can make a decision about migrating: "now that I have skills, I can look for jobs overseas, either I can apply directly or I can apply through a broker or a promoter."[60]

The language of skills and the expectation that the possession of the requisite training will enable migrants to succeed are shared by the MRC, the individuals who approach them for assistance, and the state apparatus that routinely validates the skill claims of individuals. As we have argued, however, Pakistani migrant experiences are determined by structural factors that far exceed this limited account of the problem. An in-depth examination of a typical migrant experience—such as CC's experience detailed below—reveals that although "skills" can be learned in a variety of ways, the structure of recruitment, inequality between Pakistan and the Gulf, and the racialized labor market in the Gulf continue to be the decisive factors shaping the migrant experience. The story below demonstrates a typical migration experience: Despite acquiring mastery over a specific set of skills on repeated occasions, CC's skills were constantly devalued and his experience thwarted by the immigration and labor market regimes in the Gulf.

CC's Story

> I started my training as an electrician in 2007 at a local electronics repair workshop. It was going well in the beginning but with the passage of time, continuing my training became a bit hard for me. The workshop owner who was my mentor was not teaching me the work effectively or helping me improve my skills, and instead kept me busy with mundane tasks like cleaning the workshop and preparing tea for him. He was also verbally abusive and paid me very little. So I switched jobs and started working at an air-conditioning repair shop a year later. I was there for about 2 to 3 years and learnt a lot about how to repair ACs. The workshop owner was very helpful in training me. When I became a skilled air-conditioning technician, I asked for an increase in my salary but that did not happen. Frustrated by this, in 2013, I decided to go to Dubai, hoping to be able to earn more there.[61]

Although CC did not go through formal institutional routes to acquire technical education credentials, he attempted to acquire mastery of a trade through on-the-job training and was successful. Relying on local employment opportunities where training takes place under the tutelage of a more experienced worker is a process familiar to many Pakistanis. Yet CC's remarks demonstrate that despite his proficiency in fixing air conditioners, his skill was denied and did not equate to increased earning potential. CC thus decided to migrate to the Gulf.

CC's uncle, who had been based in Saudi Arabia for a number of years, served as his trusted in-house migration expert and intermediary to Gulf employment. Like all other Gulf-bound labor migrants, CC had to wade through layers of migration bureaucracy, completing a number of essential pre-departure processes. These included getting his job contract vetted and stamped by the Protectorate of Emigrants, going through a compulsory medical checkup mandated by the UAE government, and attending a pre-departure orientation training organized by the Protectorate. Despite the fact that the Bureau purports to vet all outgoing migrants, CC was never asked specifically about his relevant skills and competencies, and was not asked to show any documents or certificates referencing his work experience or qualifications. There were no mandatory testing or training requirements that CC had to go through, and he assumed that information about his skills and experience had been conveyed to his employer informally by either his uncle or his uncle's friend. CC had spent years in Pakistan acquiring technical skills—but to what end? It was not even clear who exactly his employer would be in the Gulf, and what he would be doing when he arrived.

CC's description of the time he spent in the UAE reflects a complicated yet familiar pattern of the migration experience, one that is hard to characterize in simplistic positive or negative terms. CC flew from Peshawar to Dubai in October 2013, was greeted at the airport by a representative of the company that had hired him, and was transported along with a number of other men to company housing in Abu Dhabi. While CC had expected to begin his new job right away, to his surprise he had no work at all for the first two months, and spent his time in the workers' accommodation provided by his employer. During these months he did not receive any wages. While CC did not find daily life unpleasant, he grew increasingly concerned as the days passed without any sign of work or wages.

Almost two months after arriving in the UAE, CC was finally sent to work in an automobile workshop in Dubai. He quickly took to his new job and felt well-equipped to carry out the tasks assigned to him. The daily workload was manageable and the atmosphere collegial, but CC found that the living arrangements in Dubai left a lot to be desired. While CC struggled with his living conditions, now that he was finally working and getting paid, he was able to start sending money back home to his family. CC had a fairly contented eight months in Dubai, was looking forward to seeing his two-year contract through to completion, and, given how well things were going at work, was even hoping that his contract might be extended.

In August of 2014, a new supervisor, also a Pakistani, was hired to manage CC's workshop. Shortly after stepping into the role, the new supervisor decided that efficiency and profitability measures were needed, and that staff cuts would be imposed. To his shock, CC's contract was abruptly terminated as part of the new downsizing measures. CC knew that he had been doing well at the workshop and that this termination was not based on performance issues or any lack of ability on his part. Along with some colleagues who had also lost their contracts, CC tried to negotiate with both the supervisor and other senior staff at the company, but their attempts to convince their employer to reinstate their contracts were unsuccessful. Having exhausted the various options open to him that might have made staying on possible, CC went back to Pakistan after having spent only ten months in the UAE. He arrived home extremely frustrated that his migration journey had been brutally cut short and angry at having invested such a large amount of money for just a few months of work.

Back in Pakistan, CC sought out the OEP that had helped him with his paperwork. CC asked to be paid back at least a portion of the fees they had extracted from his uncle. To CC, it seemed unfair that the OEP should hold on to a payment that had been made for a two-year contract but translated into only eight months of earned wages. CC pointed out that his contract had been terminated without any fault on his part, as there had been nothing wrong with the quality of his work. OEP staff offered CC commiseration and sympathy but refused to return any of the fees that they had been paid. They suggested that CC try for another Gulf job and told him that they would help him find a suitable position. They would not be able to waive the recruitment fees, however, and he would have to once again front about two thousand U.S. dollars. But, for CC, once was enough. He was unwilling to risk yet more money on a venture dependent on too many imponderable factors beyond his control. CC's one and only experience at Gulf migration had convinced him that employment in the Gulf would always be highly insecure, despite his demonstrated aptitude and capacity for the work, and that at any time he could be vulnerable to losing a contract and his visa.

CC's description of his experiences—during pre-departure, his time in the Gulf, as well as upon his return—mirrors the migrant trajectories of many other Pakistanis. Their experiences are shaped by a variety of factors, including their positionality in Pakistan as well as in the Gulf. CC's inability to turn his migration into a financial success was not because of a lack of skill or inability to carry out his job, but due to broader social and economic limitations that remain central and structural features of this migration system. CC's story highlights the inherent imbalance between, and dependency of, labor migrants and their employers; migrants lack any social protection when things go wrong. While the data examined demonstrates a continuous deflection of blame and responsibility for the conditions faced by Pakistani migrants—from the Pakistani state, to the recruitment agents, to receiving states, to the workers themselves—perhaps what is clearest is the pervasiveness of the structures that maintain the subordination of the South Asian worker in the Gulf economy.

Conclusion

This chapter has presented a paradox regarding technical education and job opportunities for Pakistani labor migrants in the Gulf. Pakistani migrants aspire to be *more* than interchangeable bodies that merely have to be "physically sound" to perform the manual labor demanded by employers in the Gulf. However, the Pakistani state's technical education apparatus ends up reproducing rather than challenging the racial structure of the Gulf labor market.

Given the limited local opportunities for economic advancement, however, many Pakistanis still choose to migrate to the Gulf in the hope of gaining economic success. Interviews with migrants, government agencies, civil society actors, and labor recruiters reflect the empirical reality of the challenges migrants experience when navigating the recruitment processes and the cumbersome bureaucracies of the state in both sending and receiving countries. Many Pakistani migrants assess the best opportunities available to them and deploy their own networks and resources to access Gulf labor markets. For more than a century now, complex migrant-driven employment and recruitment networks have been established in the South Asia–Gulf migration corridor, alongside formal public and private channels of recruitment. Still, migrants' personal agency, the robustness of their networks and connections, and the trustworthiness of friends in the Gulf cannot surmount the tenuousness of their presence, the devaluation of their labor, and the racialized nature of the labor market in the Gulf.

Sensational accounts of low-wage migration to the Gulf have invited imprecise comparisons between contemporary migration regimes, indentured servitude, and slavery by emphasizing the coercive and unfree nature of these movements. In this chapter, we refocus attention away from instances of contract substitution, violence, and deceptive brokers to examine instead the role that regular and mundane institutions—such as the technical education apparatus—play in ideologically legitimizing the Gulf-Pakistan labor migration complex, obfuscating its inherent inequalities. We focus on the distinctive role that South Asian labor has come to play in fueling processes of capital accumulation globally—at one time in the West Indies and today in the Gulf. The sheer volume of people traversing this particular migration corridor should alert us to its significance as a site of importance to histories of capitalism. Thinking in line with other scholars in this volume, we position the technical education apparatus as a modern iteration of historical industrial education projects. This case demonstrates the performative role educational institutions have played in reinforcing racial hierarchies by providing ideological legitimacy to deeply exploitative capitalist structures.

Notes

1. These interviews are part of a rapid ethnographic study that was undertaken over several weeks in 2017 in Islamabad, the capital city of Pakistan, and in Peshawar, the provincial capital of Khyber Pakhtunkhwa. The use of a rapid ethnographic study allows for immersive qualitative research within a limited timeframe to maximize research efficiency, utilize inductive research methods, and ensure that the research is participatory in nature and takes local dynamics into account. All names of agencies have been anonymized for the purpose of this chapter. The insights in this paper furthermore draw on Zahra Babar's years of working on migrant labor issues with the ILO and the Center for International and Regional Studies, Georgetown University in Qatar.

2. For instance, Rina Agarwala highlights similar dynamics confronting Indian laborers investing in technical education. However, she argues that, as an emerging economic power, India has been successful in changing the image of the Indian migrant worker in recent years. Rina Agarwala, *The Migration-Development Regime: How Class Shapes Indian Emigration* (Oxford: Oxford University Press, 2022).

3. International Labour Organisation, "From Pakistan to the Gulf Region: An Analysis of Links Between Labour Markets, Skills, and the Migration Cycle," ILO, September 30, 2016, www.ilo.org/islamabad/whatwedo/publications/WCMS_532696/lang—en/index.htm.

4. Umer Nangiana, "Visiting Delegation Pitches for Pak Workforce in Qatar," *Gulf Times*, May 25, 2015, www.gulf-times.com/story/440676/Visiting-delegation-pitches-for-Pak-workforce-in-Q.

5. "Labour Migration, Skills Development and the Future of Work in the Gulf Cooperation Council (GCC) Countries," International Labour Organisation, *ILO Working Paper* (December 2017), www.ilo.org/publications/labour-migration-skills-development-and-future-work-gulf-cooperation.

6. Natasha Iskander, *Does Skill Make Us Human? Migrant Workers in 21st-Century Qatar and Beyond* (Princeton: Princeton University Press, 2021), 6.

7. See Andrea Wright, *Between Dreams and Ghosts: Indian Migration and Middle Eastern Oil* (Stanford: Stanford University Press, 2021); Robert Vitalis, *America's Kingdom: Mythmaking on the Saudi Oil Frontier* (Stanford: Stanford University Press, 2007); Toby Craig Jones, *Desert*

Kingdom: How Oil and Water Forged Modern Saudi Arabia (Cambridge: Harvard University Press, 2011); and Omar Hesham AlShehabi, "Policing Labour in Empire: The Modern Origins of the Kafala Sponsorship System in the Gulf Arab States," *British Journal of Middle Eastern Studies* 48, no. 2 (2021): 291–310.

8. Amélie Le Renard, *Western Privilege: Work, Intimacy, and Postcolonial Hierarchies in Dubai* (Stanford: Stanford University Press, 2021).

9. Ahmed Kanna, Amélie Le Renard, and Neha Vora, *Beyond Exception: New Interpretations of the Arabian Peninsula* (Ithaca: Cornell University Press, 2020).

10. ILO, "Labour Migration: Facts and Figures."

11. Mehran Kamrava and Zahra Babar, eds. *Migrant Labor in the Persian Gulf* (New York: Columbia University Press, 2012), 7–10; Zahra R. Babar, "Labor Migration to the Persian Gulf Monarchies," in *The Oxford Handbook of the Sociology of the Middle East*, ed. Armando Salvatore, Sari Hanafi, and Kieko Obuse (Oxford: Oxford University Press, 2020).

12. Nasra M. Shah, Almazia Shahzad, and Samar Quddus, *Pakistan Migration Report 2022* (CIMRAD: Lahore School of Economics, 2022), 5–9.

13. "Labour Migration Report 2020," Bureau of Emigration and Overseas Employment, Government of Pakistan, https://beoe.gov.pk/files/statistics/yearly-reports/2020/2020-full.pdf.

14. Zahra Babar, "Purveyors of Dreams: Labour Recruiters in the Pakistan to Saudi Arabia Migration Corridor," *Migration and Development* 10, no. 1 (2021), 68–70.

15. Thomas Gammeltoft-Hansen and Ninna Nyberg Sorensen, eds., *The Migration Industry and the Commercialization of International Migration* (London: Routledge, 2013); Julien Debonneville, "An Organizational Approach to the Philippine Migration Industry: Recruiting, Matching and Tailoring Migrant Domestic Workers," *Comparative Migration Studies* 9, no. 12 (2021).

16. Babar, "Purveyors of Dreams."

17. Omer Karasapan, "Pandemic Highlights the Vulnerability of Migrant Workers in the Middle East," Brookings Institute, September 17, 2020, www.brookings.edu/blog/future-development/2020/09/17/pandemic-highlights-the-vulnerability-of-migrant-workers-in-the-middle-east.

18. Gammeltoft-Hansen and Sorensen, eds. *The Migration Industry*; Debonneville, "An Organizational Approach to the Philippine Migration Industry."

19. Onur Ulas Ince, "Deprovincializing Racial Capitalism: John Crawfurd and Settler Colonialism in India," *American Political Science Review* 116, no. 1 (2021): 144–160; Inés Valdez, "Socialism and Empire: Labor Mobility, Racial Capitalism, and the Political Theory of Migration," *Political Theory* 49, no. 6 (2020): 902–933; Muriam Haleh Davis, *Markets of Civilization: Islam and Racial Capitalism in Algeria* (Durham: Duke University Press, 2022).

20. Adam Hanieh, *Money, Markets, and Monarchies: The Gulf Cooperation Council and the Political Economy of the Contemporary Middle East* (Cambridge: Cambridge University Press, 2018).

21. Hanieh, *Money, Markets, and Monarchies.*

22. Interview, PK 009, Focus Group with 40 return migrants, Chak Shehzad, Pakistan, May 2017.

23. Jonathan Ngeh and Michaela Pelican, "Intersectionality and the Labour Market in the United Arab Emirates: The Experiences of African Migrants," *Zeitschrift für Ethnologie (ZfE)/Journal of Social and Cultural Anthropology* (2018): 171–194; Rhacel Salazar Parreñas, *Servants of Globalization: Migration and Domestic Work* (Stanford: Stanford University Press, 2015).

24. Noora Lori, *Offshore Citizens: Permanent Temporary Status in the Gulf* (Cambridge: Cambridge University Press, 2019).

25. Neha Vora and Natalie Koch, "Everyday Inclusions: Rethinking Ethnocracy, Kafala, and Belonging in the Arabian Peninsula," *Studies in Ethnicity and Nationalism* 15, no. 3 (2015): 540–552.

26. In fact, in Kuwait and Qatar, they outnumber citizens about 9 to 1. Furthermore, not only has the ethno-national composition of the Gulf's migrant population changed over the past decades, so too has its gender balance (or rather its gender imbalance) as the vast majority of workers are "single" male migrant workers. International Labour Organisation, "Labour Migration: Facts and Figures," 2021, www.ilo.org/beirut/areasofwork/labour-migration/lang—en/index.htm.

27. Neha Vora and Amélie Le Renard, "Who Is 'Indian' in the Gulf? Race, Labor and Citizenship," MERIP no. 299 (2021), https://merip.org/2021/08/who-is-indian-in-the-gulf-race-labor-and-citizenship-299.

28. Bina Fernandez, "Racialised Institutional Humiliation Through the *Kafala*," *Journal of Ethnic and Migration Studies* 47, no. 19 (2021): 4344–4361; Hisham Aidi, Marc Lynch, and Zachariah Mampilly, "Racial Formations in Africa and the Middle East: A Transregional Approach," *POMEPS Studies* 44, September 2021, https://pomeps.org/wp-content/uploads/2021/09/POMEPS_Studies_44_Web-rev3.pdf. This compendium brings together a collection of short essays from a variety of scholars working across the wider Middle East and North African region, explicitly adopting the lens of racialization to approach their cases.

29. Le Renard, *Western Privilege.*

30. Lisa Lowe, *The Intimacies of Four Continents* (Durham: Duke University Press, 2015); Matthew Guterl and Christine Skwiot, "Atlantic and Pacific Crossings: Race, Empire and the 'Labor Problem' in the Late Nineteenth Century," *Radical History Review*, no. 91 (2005): 40–61.

31. Ince, "Deprovincializing Racial Capitalism."

32. Radhika Singha, *A Despotism of Law: Crime and Justice in Early Colonial India* (Oxford: Oxford University Press, 1998); Indrani Chatterjee, *Gender, Slavery and Law in Colonial India* (Delhi: Oxford University Press, 1999); Louis Dumont, *Homo Hierarchicus: The Caste System and Its Implications* (Chicago: Chicago University Press, 1980).

33. Tayyab Mahmud, "Cheaper Than a Slave: Indentured Labor, Colonialism, and Capitalism," *Whittier Law Review* 34, no. 2 (2013): 215–243.

34. Indrani Chatterjee, "Slavery, Semantics and the Sound of Silence," in *Slavery and South Asian History*, ed. Indrani Chatterjee and Richard M. Eaton (Bloomington: Indiana University Press, 2006), 287–315; Rupa Viswanath, *The Pariah Problem: Caste, Religion, and the Social in Modern India* (New York: Columbia University Press, 2014); Andrea Major, *Slavery, Abolitionism and Empire in India 1772–1843* (Liverpool: Liverpool University Press, 2012); Nira Wickramasinghe, *Slave in a Palanquin: Colonial Servitude and Resistance in Sri Lanka* (New York: Columbia University Press, 2020).

35. Radhika Mongia, *Indian Migration and Empire: A Colonial Genealogy of the Modern State* (Durham: Duke University Press, 2018), 56–84.

36. Mahdavi Kale, *Fragments of Empire: Capital, Slavery, and Indian Indentured Labor in the British Caribbean* (Philadelphia: University of Pennsylvania Press, 2010); Rachel Sturman, "Indian Indentured Labor and the History of International Rights Regimes," *American Historical Review* 119, no. 5 (2014): 1439–1465.

37. James Onley, "The Raj Reconsidered: British India's Informal Empire and Spheres of Influence in Asia and Africa," *Asian Affairs* 40, no. 1 (2009): 44–62; Andrew M. Gardner, *City*

of Strangers: Gulf Migration and the Indian Community in Bahrain (Ithaca: Cornell University Press, 2010).

38. Hideaki Suzuki, "Baluchi Experiences Under Slavery and the Slave Trade of the Gulf of Oman and the Persian Gulf, 1921–1950," *Journal of the Middle East and Africa* 4, no. 2 (2013): 205–223.

39. Johan Mathew, *Margins of the Market: Trafficking and Capitalism Across the Arabian Sea* (Oakland: University of California Press, 2016). As Johan Mathew shows, abolishing the slave trade in the British Empire ended a particular form of trade in humans ("slavery"), but normalized a whole range of other potentially coercive "free" migrations facilitated by the colonial state.

40. Andrea Wright, "Slaves to Contract Workers: Genealogies of Consent and Security in Indian Labor Migration," *Journal of World History* 32, no. 1 (2021): 29–41; Agarwala, *The Migration Development Regime*, especially Chapter 3, "The Rise and Fall of the Coolie MDR."

41. These key texts represent a few of the major articulations of the main tenets of the racial capitalism framework, although there have been many more over the past several years: Cedric J. Robinson, *Black Marxism: The Making of the Black Radical Tradition* (Chapel Hill: University of North Carolina Press, 2000); Destin Jenkins and Justin Leroy, eds., *Histories of Racial Capitalism* (New York: Columbia University Press, 2021); Gargi Bhattacharyya, *Rethinking Racial Capitalism: Questions of Reproduction and Survival* (London: Rowman & Littlefield, 2018); Michael Ralph and Maya Singhal, "Racial Capitalism," *Theory and Society* 48, no. 6 (2019): 851–881; Jodi Melamed, "Racial Capitalism," *Critical Ethnic Studies* 1, no. 1 (2015): 76–85; Robin D. G. Kelley, "What Did Cedric Robinson Mean by Racial Capitalism?" *Boston Review*, January 12, 2017, www.bostonreview.net/articles/robin-d-g-kelley-introduction-race-capitalism-justice.

42. Zahra Babar, "Introduction," in *Arab Migrant Communities in the GCC*, ed. Zahra Babar (Oxford: Oxford University Press/Hurst, 2017), 1–9.

43. Zahra Babar, "Introduction," 1–9.

44. Natasha Ridge, Soha Shami, and Susan Kippels, "Arab Migrant Teachers in the United Arab Emirates and Qatar: Challenges and Opportunities," in *Arab Migrant Communities in the GCC*, ed. Zahra Babar (Oxford: Oxford University Press/Hurst, 2017), 40–43; Abbie Taylor, Nada Soudy, and Susan Martin, "The Egyptian 'Invasion' of Kuwait: Navigating Possibilities Among the Impossible," in *Arab Migrant Communities in the GCC*, ed. Zahra Babar (Oxford: Oxford University Press/Hurst, 2017), 90–93.

45. Iskander, *Does Skill Make Us Human?*, 41–51.

46. Helene Thiollet, "Migration as Diplomacy: Labor Migrants, Refugees and Arab Regional Politics in the Oil-Rich Countries," *International Labor and Working-Class History* (2011): 3–6.

47. Ridge, Shami, and Kippels, "Arab Migrant Teachers in the United Arab Emirates and Qatar," 43.

48. Vora, *Impossible Citizens*.

49. For some of the literature on this see: Claire Beaugrand, *Stateless in the Gulf: Migration, Nationality, and Society in Kuwait* (London/New York: I. B. Taurus, 2018); Anh Nga Longva, "Neither Autocracy nor Democracy but Ethnocracy: Citizens, Expatriates and the Socio-Political System in Kuwait," in *Monarchies and Nations: Globalisation and Identity in the Arab States of the Gulf*, ed. Paul Dresch and James Piscatori (London: I. B. Tauris 2005), 96–114; and Zahra Babar, "Enduring 'Contested' Citizenship in the Gulf Cooperation Council," in *The Middle East in Transition: The Centrality of Citizenship*, ed. Nils A. Butchenson and Roel Meijer (Cheltenham: Edward Elgar, 2018), 115–132.

50. Lori, *Offshore Citizens*, 133–159.

51. Philippe Fargues, "Immigration Without Inclusion: Non-Nationals in Nation-Building in the Gulf States," *Asian and Pacific Migration Journal* 20, no. 3–4 (2011): 277; Martin Baldwin-Edwards, "Labour Immigration and Labour Patterns in the GCC Countries: National Patterns and Trends," Kuwait Programme on Development, Governance and Globalisation in the Gulf States, *Research Paper* no. 15 (2011): 7–10.

52. Babar Shahbaz, Qasim Ali Shah, Abid Q. Suleri, Steve Commins, and Akbar Ali Malik, "Livelihoods, Basic Services, and Social Protection in North-Western Pakistan, Sustainable Development Policy Institute, *Working Paper* 5, August 2021, https://luskin.ucla.edu/sites/default/files/downloadpdfs/Livelihoods,%20basic%20services%20and%20social%20protection%20in%20north-western%20Pakistan%20(1).pdf; and Z. U. Ahmed, C. Euler, S. G. Khattak, J. F. Morton, and M. Tariq, "Assessing the Impact of Community-Driven Development: 26 years of Pakistan's Rural Support Programmes," Rural Support Programmes Network, 2009, www.rspn.org/wp-content/uploads/2013/04/Assessing-the-Impact-of-Community-Driven-Development-Twenty-Six-Years-of-Pakistans-Rural-Support-Programmes-2009.pdf.

53. "A Migrant's Journey for Better Opportunities: The Case of Pakistan," World Bank, 2019, https://documents1.worldbank.org/curated/en/540841530861637430/pdf/Pakistan-International-Mobility-Draft-06.pdf.

54. "Pakistan Needs More Vocational Training," *Dawn*, February 3, 2020, www.dawn.com/news/1532054.

55. Overseas employment promoter or OEP is the terminology used in Pakistan to refer to private, licensed labor recruiters.

56. PK 16, Anonymized interview with labor recruiter, May 2017, Rawalpindi, Pakistan.

57. PK 05, Anonymized interview with labor recruiter, May 9, 2017, Peshawar, Pakistan.

58. PK 11, Non-anonymized interview with Director General Kashif Ahmed Noor, Bureau of Emigration and Overseas Employment, May 15, 2017, Islamabad, Pakistan.

59. The Protectorate of Emigrants is the provincial branch of the Bureau of Emigration and Overseas Employment under Pakistan's Ministry for Labour, Manpower and Overseas Pakistanis. Eight Protectorates have been established across the country: one in each of the four provincial capitals, with the remaining four located in parts of the country that see the highest numbers of out-migration.

60. PK 10, Non-anonymized interview with Kayyam Sohail, Migrant Resource Center, May 15, 2017, Islamabad, Pakistan.

61. PK 18, Anonymized interview with CC, a Pakistani return migrant, August 2, 2017, Islamabad, Pakistan.

CONTRIBUTORS

Lukas Allemann has a background in oral history and Russian studies. He has been working in the anthropology research team of the Arctic Centre, University of Lapland, since 2013. He earned his PhD degree in 2020, with an inquiry about the Soviet policies toward Indigenous minorities from a grassroots perspective, based on research among Russia's Sámi people. Another strand of research has been anthropological inquiry into young people's aspirations and well-being in Arctic single-industry towns. Methods of oral history, participant observation, archival work, and media analysis are equally represented in Lukas's work. In all of his research, Lukas puts an emphasis on long-term field commitment and bringing back research results to the communities they stem from.

Danya Al-Saleh is an Assistant Professor in International Studies at the University of Washington. She is a geographer with research and teaching interests in political economy, energy transition, environmental and climate justice, critical university studies, and the Middle East/North Africa. Her current book project explores the transnational everyday and institutional politics of energy transition and contributes to studies of fossil-fueled capitalism, education, engineering, and gender. She has published in *Environment and Planning E: Nature and Space* (*EPE*), *Climate and Development*, *International Journal of Middle Eastern Studies*, *Environment and Planning C: Politics and Space*, and *Gender, Place, and Culture*.

Hossein Ayazi is Senior Policy Analyst at the Othering & Belonging Institute at the University of California, Berkeley. Across his research, teaching, and policy work, he engages the U.S. and global political economy of race, agri-food systems, and environmental change, and their relationship to antiracist, anticolonial, and revolutionary-socialist struggles from the twentieth century on into the present. Previously, he was a Postdoctoral Fellow and Visiting

Scholar at Tufts University and a Visiting Assistant Professor at Williams College. He holds a PhD in Environmental Science, Policy, and Management from UC Berkeley.

Zahra Babar is Executive Director of the Center for International and Regional Studies (CIRS) at Georgetown University in Qatar. Her recent publications include: "The Buying of Freedom: Migrant Workers and the Azad Visa in the Gulf," *International Migration* (2024); "The 2022 World Cup and Migrants' Rights in Qatar: Racialised Labour Hierarchies and the Influence of Racial Capitalism," with N. Vora, *The Political Quarterly* (2022); "The Vagaries of the In-Between: Labor Citizenship in the Persian Gulf," *International Journal of Middle East Studies* (2020); "Purveyors of Dreams: Labour Recruiters in the Pakistan to Saudi Arabia Migration Corridor," *Migration and Development* (2020); and "Qatar, the World Cup, and the Global Campaign for Migrant Workers' Rights," in *Football in the Middle East: State Society and the Beautiful Game* (Hurst/Oxford University Press, 2022). Her most recent book is a co-edited volume titled *Transnational Generations in the Arab Gulf States and Beyond* (Springer, 2024).

Oli Charbonneau is Lecturer in American History at the University of Glasgow, where they research and teach histories of American empire. Charbonneau's publications include: *Civilizational Imperatives: Americans, Moros, and the Colonial World* (Cornell University Press, 2020), which explores a half-century of American colonial rule in the Philippines' Muslim South; articles in *Diplomatic History*; *Modern American History*; *The Journal of the Gilded Age and Progressive Era*; and a chapter in *Crossing Empires: Taking U.S. History into Transimperial Terrain* (Duke University Press, 2020). They are currently completing a global microhistory of the Lake Mohonk Conference of Friends of the Indian and Other Dependent Peoples that is supported by fellowships from the Royal Society of Edinburgh and Haverford College. They are the recipient of the 2022 Stuart L. Bernath Lecture Prize from the Society for Historians of American Foreign Relations.

Bronwen Everill (FRHistS) is the author of *Africonomics: A History of Western Ignorance* (2024); *Not Made by Slaves: Ethical Capitalism in the Age of Abolition* (2020); and *Abolition and Empire in Sierra Leone and Liberia* (2013). She has been the recipient of a Leverhulme Fellowship and her research has been supported by the British Academy, Economic History Society, Huntington

Library, John Carter Brown Library, Library Company of Philadelphia, and the Cambridge-Africa Alborada Fund. She is on the faculty of the Princeton Writing Program and is a research affiliate at the Laboratory for the Economics of Africa's Past, Stellenbosch University.

Mishal Khan is a historical sociologist examining legacies of slavery and abolition, the intersections of race and capitalism, and global political economy and empire. Mishal's work leverages research on historical forms of labor subordination to critically analyze contemporary forms of labor precarity and exploitation in the U.S. and global economy. Mishal has worked on technology, platform work, trafficking, and labor migration issues and has held postdoctoral fellowships at The University of Texas at Austin; the University of California College of the Law, San Francisco; and the Gilder Lehrman Center for the Study of Slavery, Resistance, and Abolition at Yale University. Mishal has a degree in sociology from the University of Chicago and is currently a senior researcher at the University of California Berkeley Labor Center.

Arun Kumar is a historian of modern India and an Assistant Professor of British Imperial, Colonial, and Postcolonial History at the University of Nottingham. His work explores various facets of working-class life histories, including their dreams, education, nighttime histories, sleep, and literary culture. Kumar's research has appeared in journals such as *Past & Present*; *South Asia: Journal of South Asia*; *Journal of South Asian Development*; and *International Review of Social History*; and his public-facing writings have been translated into Ukrainian and Russian.

Janne Lahti is a historian working as a researcher at Linnaeus University. He is also an Associate Professor (title of docent) at the University of Helsinki. He researches global and transnational histories of settler colonialism, the American West, German colonialism, and Nordic colonialism. He has published numerous articles and nine books, including *Colonial Violence and Monuments in Global History* (with Cynthia C. Prescott, Routledge 2023); *Finnish Settler Colonialism in North America* (with Rani-Henrik Andersson; Helsinki University Press, 2022); *German and United States Colonialism in a Connected World* (Palgrave, 2021); and *The American West and the World* (Routledge, 2019). Lahti also serves as chief editor of the journal *Settler Colonial Studies*.

Dolf-Alexander Neuhaus is senior research fellow at the German Institute for Japanese Studies in Tōkyō. His research focuses on the modern history of Japan and Korea and their transregional entanglements with Asia, the history of education and Christianity in Korea and Japan, and the dissemination and adaptation of technical education and assistance in Asia within the framework of Japanese official development aid during the Cold War. His recent publications include *Beyond Nation and Empire: Interactions Between Korean Students and Japanese Protestants (1880–1923)* (2024, in German) and an article on the encounter of early Korean converts with Protestantism in Japan: "El encuentro del protestantismo japonés con la Corea antes de la primera guerra sino-japonesa," in *Revista Internacional de Estudios Asiáticos* (2024, in Spanish). Neuhaus has guest-edited (with Klaus Dittrich) a special issue on "'Education Fever'? Korea from the Late Nineteenth to the Early Twenty-First Century," in *History of Education* (2023), which also includes his article "Beyond Assimilation and National Resistance: 'Education Fever' During the Colonial Period in Korea, 1910–1945," *History of Education* (2023).

Sarah Steinbock-Pratt is an Associate Professor of History at the University of Alabama. She is the author of *Educating the Empire: American Teachers and Contested Colonization in the Philippines* (Cambridge University Press, 2019). Steinbock-Pratt is currently working on a second book project, a microhistory of a Black-Filipino family, which examines transimperial migration and meanings of race in the early twentieth century.

Karine V. Walther is an Associate Professor of History at Georgetown University in Qatar. She holds an MA, MPhil, and PhD in History from Columbia University, a Maîtrise and Licence in Sociology from the University of Paris VIII, and a BA from the American Studies Honors Program at The University of Texas at Austin. She is currently working on her second book: *Spreading the Faith: American Missionaries, ARAMCO and the Birth of the US-Saudi Special Relationship, 1890–1955*, forthcoming with University of North Carolina Press. Her first book, *Sacred Interests: The United States and the Islamic World, 1821–1921*, was published by UNC Press in 2015. Reviews of *Sacred Interests* can be found in the *Times Literary Supplement*, *Diplomatic History*, the *American Historical Review*, the *Journal of American History*, and other academic journals. She is currently working on an edited volume focusing on global histories and practices of Islamophobia (co-edited with Abdullah Al-Arian). Before joining Georgetown, Karine served as a visiting professor at

Middlebury College, a postdoctoral fellow at the Belfer Center for Science and International Affairs, and a visiting lecturer at the Harvard Kennedy School.

Helge Wendt is research coordinator of the "Deliberation Agora" project at the University of Jena, jointly initiated with the MPI for Geoanthropology. He studied Ancient American Studies and History at the FU Berlin and received a doctorate in Modern History from the University of Mannheim. Wendt has held research positions at the MPI for the History of Science (Berlin), including in the project on the history of the Anthropocene and at the CRC Episteme in Motion. He served as a visiting professor of Latin American history at the Universities of Bielefeld and Münster and of the history of science at the University of Jena. Wendt is the author of the monographs *Die Missionarische Gesellschaft* (Steiner Verlag, 2011); *Geschichte des mestizischen Europas* (Springer, 2019); and *Kohlezeit* (Campus, 2023), as well as various collected works on topics related to the global history of knowledge.

Christine Whyte is a Senior Lecturer in Global History at the University of Glasgow, Scotland. She served on the Steering Committee for the first Care Experienced People History Month in Scotland, coordinated by Who Cares? Scotland. Her research focuses on the history of children and families in the British Empire, slavery and abolition in West Africa, and the suppression of the slave trade in the Indian and Atlantic Oceans.

INDEX

ACKNOWLEDGMENTS

Our gratitude to the contributors, each of whom expanded our notions of how to think about labor pedagogy in their own unique ways. They were a terrific group to collaborate with. Thanks to everyone at the University of Pennsylvania Press: Chris Dietrich and the series editors for Power, Politics, and the World, who saw the value of interrogating industrial education (and its legacies!) in a global context; Bob Lockhart, who, alongside Chris, gave the volume constructive, critical feedback at every stage, strengthening it; the anonymous reviewers who recognized oversights we ourselves could not see. Above all, we wish to thank the team at the Center for International and Regional Studies (CIRS), Georgetown University in Qatar: Zahra Babar, Elizabeth Wanucha, and Misba Bhatti provided incredible support during the workshop process and oriented our online collaboration in the early days of the pandemic. Their efforts before, during, and after the workshops helped the project find its rhythm quickly. Suzi Mirgani stepped in after that and ensured the wheels didn't come off, serving alternately as project coordinator, copy editor, or investigator (depending on the day); the editors and contributors owe her many debts—we could not have done this without her brilliant interventions. Oli Charbonneau wishes to thank their brilliant colleagues and co-conspirators at the University of Glasgow, who provide support, laughter, and gentle provocation; Augie, sofa-mate extraordinaire and assistant editor; Victoria, who tolerated long digressions on nineteenth-century pedagogy; and Nijey, who makes everything brighter. Karine V. Walther wishes to thank members of WAW, notably Noha Aboueldahab, Vicky Googasian, Trish Kahle, and Nadya Sbaiti for their ongoing support as amazing colleagues and the best of friends, and for keeping her sane. She would also like to thank Julien and Mia for reminding her that there is a life outside of academia.

Kolibri 360 GmbH
Ferdinand-Jühlke-Straße 7
99095 Erfurt, Deutschland
produktsicherheit@kolibri360.de